endorsed for
edexcel

REVISE EDEXCEL AS/A LEVEL
Physics

REVISION GUIDE

Series Consultant: Harry Smith

Authors: Steve Woolley and Steve Adams

- -

For the full range of Pearson revision titles across KS2, KS3, GCSE, AS/A Level and BTEC visit:
www.pearsonschools.co.uk/revise

ALWAYS LEARNING

PEARSON

Contents

. . . .

A small bit of small print

Edexcel publishes Sample Assessment Material and the Specification on its website. This is the official content and this book should be used in conjunction with it. The questions in *Now try this* have been written to help you practise every topic in the book. Remember: the real exam questions may not look like this.

S.I. units

The **S.I.** (Système International d'Unités) is a globally agreed system of units, with seven base quantities.

Base units

mass	kg
length	m
time	s
current	A
temperature	K
amount of substance	mol
luminous intensity	cd

These are the base quantities of S.I. with their units. You do not need to learn about luminous intensity.

This set of base units is sufficient for all measurements that we need to make in science.

With the exception of the kilogram, base units are precisely defined in a way that can be replicated in any suitably equipped physics laboratory. At the moment the kilogram is specified by an object, the international standard kilogram, kept in Paris. All other masses are ultimately compared with this.

Derived units for mechanics

All quantities that we will meet in mechanics can be expressed in terms of mass, length and time. All other quantities are measured in terms of the base units; for example, speed is measured in $m\,s^{-1}$. Some of these derived units have their own names; for example force is measured in $kg\,m\,s^{-2}$, called newtons (N), and energy in $kg\,m^2\,s^{-2}$, called joules (J).

Derived units for electricity and thermodynamics

Electricity requires a base electrical unit, the ampere (A). The units of other electrical quantities are derived from the base units of mass, length, time and current. For example, potential difference is measured in volts, or $kg\,m^2\,s^{-3}\,A^{-1}$.

Thermodynamics requires two further base units, the kelvin (K) for temperature and the mole (mol) for amount of substance.

A tool for checking equations

Equations must be consistent in terms of units. This means that the units on both sides of the equation must be the same. If this is not the case then the equation *cannot* be correct. This fact gives you a useful tool for checking your work when you rearrange or derive equations.

$$s = \qquad ut \qquad + \tfrac{1}{2}at^2$$
$$m \quad m\,s^{-1} \times s \quad m\,s^{-2} \times s^2$$

However, consistent units do not mean an equation *is* correct:

$$v^2 = u^2 + as \quad ✗$$

This has consistent units throughout ($m^2\,s^{-2}$), but the equation is incorrect.

Worked example

Express the unit of power, the watt (W), in base units. **(4 marks)**

Use an equation relating power to other quantities.

$$P = \frac{W}{t} = \frac{Fs}{t} = \frac{mas}{t}$$

Hence the watt has the base units:
- m, mass, a base quantity, unit: kilogram, kg
- a, acceleration, has the derived unit $m\,s^{-2}$
- s, distance, a base quantity, unit: metre, m
- t, time, a base quantity, unit: second, s

so, in base units, $W \equiv kg\,m\,s^{-2}\,m\,s^{-1} \equiv kg\,m^2\,s^{-3}$
(kilogram metre squared per second cubed)

Now try this

Numerical values ($\tfrac{1}{2}$, 2, etc) have no units.

1 The work done by applying a force F through a distance s is given by $W = F \times s$. Express the S.I. unit of work, the joule, in terms of base units. **(2 marks)**

2 Show that kinetic energy (given by $E_k = \tfrac{1}{2}mv^2$) and potential energy (given by $E_p = mgh$) have the same base units. **(3 marks)**

3 Show that the *SUVAT* equation $s = \tfrac{1}{2}(u + v)t$ is valid in terms of units. For more about SUVAT see page 4. **(3 marks)**

Practical skills

Practical skills Physicists test hypotheses by experiments. A good experiment has clear objectives and is designed to give reliable measurements.

Example: What factors affect the period of a simple pendulum?

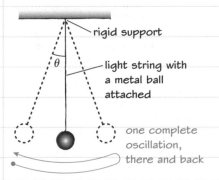

rigid support

light string with a metal ball attached

θ

one complete oscillation, there and back

The period of a simple pendulum is the time it takes for one complete oscillation. You might think that this time may depend on any or all of the following factors:

- the length of the string
- the mass of the ball
- the angle of the swing
- where the experiment is performed.

Each factor will need to be measured and varied and the effect on the period examined. In any experiment it is important to vary just one factor at a time and keep all others exactly the same.

Taking measurements

A metre rule is suitable for the **length** of the pendulum string, but the length of the pendulum is measured from the support to the centre of the ball. Measuring the diameter of the ball should be done with digital callipers or a micrometer (small lengths need to be measured with an instrument that has a small resolution). **Mass** is measured on an electronic balance; in this case, to the nearest gram is enough. Check that the balance is correctly zeroed first. If it is not, you will introduce a **systematic error**. This is an error that introduces the same error to all measurements. A protractor will give a reasonable measurement of the **angle**, within 1 or 2 degrees. The small errors in protractor readings are **random errors**. Readings might be too large or too small, by different amounts. If greater precision is required, you might measure the horizontal displacement and use trigonometry to calculate the angle.

Measuring **time** using a hand-held stopwatch introduces reaction-time errors. To minimise these, when measuring the period of a pendulum measure a convenient multiple and divide by the number of oscillations to obtain a more accurate value. Alternatively use an electronic timing system triggered by light gates.

Wherever possible, repeat measurements and take an average value to identify and reduce errors. Discard obvious outliers. Perhaps you timed nine oscillations instead of ten.

Use graphs

Remember that errors accumulate when you use several measured values in a calculation. In the pendulum experiment you are making measurements to discover which factor(s) affect the period. Presenting your results in the form of a **graph** will show a correlation more clearly and also help you to spot measurement error.

Worked example

The time constant, τ (in s), for charging a capacitor of capacitance C farads through a resistor of resistance R ohms is given by $\tau = RC$. What is the range of possible values for τ if $R = 100\ \text{k}\Omega \pm 5\%$ and $C = 10\ \mu\text{F} \pm 10\%$? **(4 marks)**

R may range from 0.95×10^5 to $1.05 \times 10^5\ \Omega$, C may range from 0.9×10^{-5} to 1.1×10^{-5} F. Thus τ will range from
$(0.95 \times 10^5\ \Omega) \times (0.9 \times 10^{-5}\ \text{F})$
$= 0.855\ \text{s}$ to $(1.05 \times 10^5\ \Omega) \times (1.1 \times 10^{-5}\ \text{F})$
$= 1.155$
The % error of τ is $\pm 15\%$.

Now try this

1 (a) Name one cause of error when using a hand-operated stopwatch. **(1 mark)**
   (b) Describe how the effects of this type of error can be reduced. **(2 marks)**
   (c) Compare the effect of this timing error on the measurement uncertainty of short and long time intervals. **(2 marks)**

2 Explain how graphs help to identify measurement errors. **(2 marks)**

3 Suggest suitable measuring instruments for the following measurements:
   (a) the dimensions of a classroom **(1 mark)**
   (b) the height of a bench above the floor **(1 mark)**
   (c) the thickness of a human hair. **(1 mark)**

Estimation

With practice, you can estimate the size of the answer before you carry out a calculation in physics.

Estimation and orders of magnitude

If you bought a bottle of juice and a packet of crisps and were charged £350, you would know someone had made a mistake. You can develop a similar 'feel' for the likely sizes of answers to physics questions, and should be able to make reasonable estimates of the physics quantities you will encounter in this course.

One way to check your answers is to do a rough calculation of the **order of magnitude** of the answer before you do the full calculation.

Quantity	Order of magnitude
nuclear radius	10^{-15} m
atomic radius	10^{-10} m
human height	10^{0} m
Earth radius	10^{7} m
Earth orbit	10^{11} m

Fermi questions

A good way to develop your skills in estimation is to practise Fermi questions. These are very rough 'back of the envelope' calculations in which you have to estimate the starting information. Famously, the nuclear physicist Enrico Fermi asked his students 'How many piano tuners are there in Chicago?', giving them only the population of Chicago as a starting point.

 1. Identify the relevant physical quantities.

2. Estimate the value of each quantity using your knowledge (not a guess!).

3. Combine the quantities.

Worked example

Estimate the number of atoms in a house brick. **(3 marks)**

Estimate brick volume:
30 cm × 15 cm × 10 cm = 4500 cm³

This is an estimate only so round to
5000 cm³ = 5×10^{-3} m³.

Estimate atom volume: $(10^{-10}$ m$)^3 = 10^{-30}$ m³

number of atoms $N = \dfrac{\text{volume of brick}}{\text{volume of atom}}$

$N = \dfrac{5 \times 10^{-3}}{10^{-30}} = 5 \times 10^{27}$ atoms

Write down your estimates for the volume of the brick and the volume of one atom.

Worked example

Estimate to one significant figure:
- the mass of an adult man and adult woman
- the volume of a classroom
- the mass of a plank of wood 2 m long. **(3 marks)**

Mass of a human: use your own mass as a starting point – answers of ~80 kg and 60 kg

Volume of a classroom: estimate length, breadth and height – answer around
10 × 10 × 3 = 300 m³

Mass of a plank: think how easy it is to lift a plank – answer ~5 kg.

Comparisons

You can sense-check some answers by comparison with things you already know. How much force is 10 N? You know that weight $W = mg$ and g (the acceleration of free fall) is $9.81\,\text{m s}^{-2}$, or roughly $10\,\text{m s}^{-2}$. So you exert about 10 N to lift up a 1 kg pack of sugar or flour.

Sense check: wood floats in water, which has a density of $1000\,\text{kg m}^{-3}$, so its density must be of that order (in fact, slightly less).
Volume of plank ~$(2 \times 0.1 \times 0.02)$ m³ so density
~5/0.004 = $1000\,\text{kg m}^{-3}$ to 1 s.f.

Now try this

1 Estimate the order of magnitude of: the mass of a car, an atom, an Earth-sized planet. **(3 marks)**

2 Estimate to one significant figure the pressure ($p = F/A$) exerted by an elephant's feet and by the wheels of a passenger aircraft. **(4 marks)**

SUVAT equations

The SUVAT equations describe the motion of bodies moving with constant (uniform) acceleration. They are sometimes called the **kinematic equations of motion**.

Variables in the SUVAT equations

s is **displacement** —— Distance travelled in a **specific direction** from a **starting point**. If the distance is in the opposite direction the displacement will be negative.

u is **initial velocity** —— **Starting** and **finishing** speeds in the specified direction. Objects travelling in the

v is **final velocity** —— other direction have negative values of velocity.

a is **acceleration** —— **Positive** values of a mean the velocity is increasing in the specified direction. **Negative** values of a mean that the velocity is decreasing in that direction – the object is decelerating, or slowing down.

t is **time** —— From the start to the end of the motion.

> Unless you are told otherwise assume that there is no air resistance or wind in questions of this type.

Four equations to learn

You should learn these four SUVAT equations:

1 $\quad v = u + at$

3 $\quad v^2 = u^2 + 2as$

2 $\quad s = ut + \frac{1}{2}at^2$

4 $\quad s = \frac{(u + v)t}{2}$

> These equations appear in the formula booklet, but it's a good idea to learn them anyway.

Worked example

A stone is released from rest at the top of a well. It hits the surface of the water after exactly 3.00 seconds. Calculate the distance between the top of the well and the surface of the water.
$(g = 9.81 \, \text{m s}^{-2})$ **(3 marks)**

$s = ? \, \text{m}, \, u = 0 \, \text{m s}^{-1}, \, a = 9.81 \, \text{m s}^{-2}, \, t = 3 \, \text{s}.$

$s = ut + \frac{1}{2}at^2$

$\quad = 0 \times 3 + \frac{1}{2} \times 9.81 \times (3)^2$

$\quad = 44.2 \, \text{m}$ (3 s.f.)

> Write down any of the SUVAT values that you know. The stone is released **from rest** so $u = 0 \, \text{m s}^{-1}$. It is free-falling under gravity so $a = 9.81 \, \text{m s}^{-2}$. You want to know s so write a question mark next to that.

> Use the equation that doesn't involve v. Always write out the equation in full before you substitute any values.

> 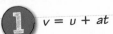 **Maths skills** The value of g is given to 3 significant figures, so you should round your answer to the same degree of accuracy.

> You need to decide whether up or down is positive. Usually up is taken as positive. Then the downward acceleration due to gravity will be negative.

Now try this

1 A car travelling at $20 \, \text{m s}^{-1}$ accelerates at a constant rate for $10 \, \text{s}$ reaching a speed of $30 \, \text{m s}^{-1}$.   **Maths skills**

   (a) Calculate how far the car travels during the period of acceleration. **(3 marks)**

   (b) Calculate the rate of acceleration of the car. **(3 marks)**

2 A catapult fires a pellet vertically upward with an initial velocity of $80 \, \text{m s}^{-1}$.

   (a) Calculate how high the pellet travels before starting to fall back to the ground. **(3 marks)**

   (b) Calculate how long the pellet will take to fall back to the ground. **(3 marks)**

Displacement–time, velocity–time and acceleration–time graphs

Velocity is the rate of change of displacement over time, and acceleration is the rate of change of velocity over time. These quantities can be related in the form of graphs.

Displacement–time (s–t) graphs

A negative slope would mean that the displacement is getting smaller, so the body is moving back in the opposite direction.

A horizontal line on an s–t graph would mean the displacement is not changing, so the body is stationary, $v = 0$.

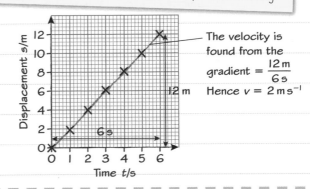

The s–t graph for the motion of a person walking

The velocity is found from the gradient $= \dfrac{12 \, m}{6 \, s}$

Hence $v = 2 \, m s^{-1}$

Velocity–time (v–t) graphs

In a plot of velocity against time, the gradient is the rate of change of velocity with time – the acceleration.

This part matches the s–t graph above: velocity is constant at $2 \, m s^{-1}$. The gradient is 0 – there is no acceleration.

The area under a v–t graph is the displacement.

The v–t graph for the motion of the person walking

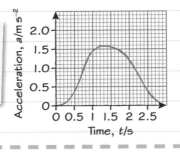

Now the velocity is decreasing – the person is slowing down and eventually stops. The acceleration (gradient) is

$$\dfrac{2.0}{2.1} = -0.95 \, m s^{-2}.$$

Acceleration–time (a–t) graphs

We can also plot acceleration against time. In most cases you will meet, acceleration is constant (straight line, gradient = 0). You may see a graph for non-uniform acceleration.

A bicycle accelerating from a standing start before settling at a constant velocity.

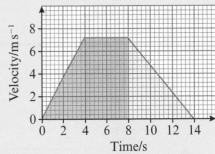

Worked example

(a) Describe the motion of the object in the velocity–time graph shown below.   **(6 marks)**

In the first 4 s the object accelerates to $7.2 \, m s^{-1}$ with acceleration – from the gradient: $1.8 \, m s^{-2}$.

For the next 4 s the object moves at a constant velocity of $7.2 \, m s^{-1}$.

In the last 6 s the object decelerates; $a = -1.2 \, m s^{-2}$.

(b) How far does the object travel?   **(4 marks)**

Total displacement = area under graph
= 14.4 + 28.8 + 21.6 = 64.8 m

The total displacement of the object is the area under the graph, found as shown in the diagram.

Now try this

1 State what the following represent.
   (a) The slope of a s–t graph, (b) the area beneath a v–t graph, (c) the slope of a v–t graph.   **(3 marks)**

2 A falling object has a constant acceleration if we assume air resistance can be ignored. In practice falling objects reach a terminal velocity. Sketch the v–t graph for the latter case.   **(3 marks)**

Scalars and vectors

Some quantities in physics have both magnitude (size) and direction. These are called vector quantities.

Scalar or vector?

Scalar quantities do not require a direction: if asked how tall you are you would not say 1.6 m up! If asked how much petrol you wanted in your car a number of litres is all you need to answer.

However, if you were giving instructions on a map, distances without direction would not be helpful. **Vector** quantities must include a direction as well as a number.

Examples from mechanics

✓ Scalar quantities: length, area, volume, mass, time, distance, speed, density, work, energy.

✓ Vector quantities: displacement, velocity, weight, acceleration, force, momentum.

Scalars and vectors on a running track

Most athletics tracks are ovals with a distance of 400 m per lap. The straights are the same length as the curves (100 m).

A child runs one lap in 80 s. Consider the distances and displacements of the child as he runs from A to B, C and D.

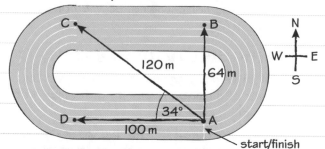

From A to ...	Distance	Displacement
B	100 m	64 m north (0°)
C	200 m	120 m 34°
D	300 m	100 m west
A	400 m	0 m

See how taking the direction into account changes things!

Worked example

Two forces are applied to an object: F_1 = 100 N in an easterly direction and F_2 = 40 N acting south. Represent them on a vector diagram. **(2 marks)**

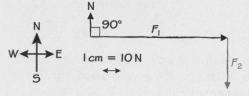

Vector quantities are often represented by arrows whose length is proportional to the size of the quantity and whose direction shows the direction of the vector. Vector diagrams must have a scale and a reference direction.

Vectors are frequently written in bold italic letters or with an arrow or line across the letter: $\boldsymbol{F}$ or $\vec{F}$

Now try this

1 State whether air pressure is a vector or scalar quantity. Give a reason for your answer. **(2 marks)**

2 The diagram shows a runner on a 400 m track. She completes one lap in 55.9 seconds.

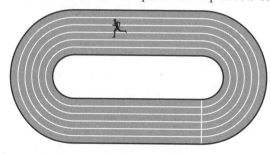

(a) What is her average speed for one lap? **(1 mark)**

(b) What is her average velocity for one lap? **(1 mark)**

Resolution of vectors

You need to be able to resolve vectors into two perpendicular components by maths or scale drawing.

Two methods for resolving vectors

You can use **trigonometry** or **scale drawing** to **resolve** vectors, like velocity or force, into two **perpendicular components**.

For example, the weight of a book on a slope resolves into two components, one holding the book flat on the surface and one pulling it downwards parallel to the slope.

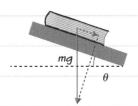

mg

θ

The vector you are resolving is always the **hypotenuse** of the vector triangle.

This is the same as the angle of the slope to the horizontal.

$mg \sin \theta$ —— This is the component of the weight that acts **parallel** to the slope. Always use **sin** for the component **opposite** the angle.

mg

$mg \cos \theta$

This is the component of the weight that acts **perpendicular** to the slope. Always use **cos** for the component **adjacent** to the angle.

Even if you are not making a scale drawing, it helps to draw a sketch of **vector triangle**.

Worked example

The diagram shows a cannonball leaving a cannon at a speed of $58 \, \text{m s}^{-1}$. The angle of launch is $25°$.

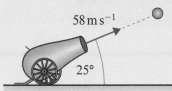

$58 \, \text{m s}^{-1}$

$25°$

Find by scale drawing:

(i) the magnitude of the horizontal component of the cannonball's velocity. **(1 mark)**

$1 \, \text{cm} = 10 \, \text{m s}^{-1}$

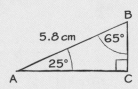

B

$5.8 \, \text{cm}$ $65°$

$25°$

A   C

$AC = 5.3 \, \text{cm}$

Horizontal component = $53 \, \text{m s}^{-1}$ (2 s.f.)

(ii) the magnitude of the vertical component of the cannonball's velocity. **(2 marks)**

$BC = 2.5 \, \text{cm}$

Vertical component = $25 \, \text{m s}^{-1}$ (2 s.f.)

Maths skills   Take a **protractor**, a **ruler** and a **sharp pencil** into your exam.

Follow these steps:

1. Draw an angle of 25° at A.

2. Measure 5.8 cm along one side and mark B.

3. Measure an angle of 90 − 25 = 65° at B.

4. Draw the line BC at this angle down to the horizontal to complete the right angled triangle ABC.

5. Measure AC to the nearest mm.

6. Use your scale to write the components.

If the question does not specify the method, you could use trigonometry:
(i) $58 \cos 25° = 52.5 \, \text{m s}^{-1}$
(ii) $58 \sin 25° = 24.5 \, \text{m s}^{-1}$

Check it! Your answer should make sense. The angle is less than 45° so the horizontal component should be greater than the vertical component.

Now try this

This diagram shows a painting hanging on a wall from two wires.
Wire A has a tension $T_A = 87.5 \, \text{N}$ and is at 52° to the horizontal, and wire B has a tension $T_B = 62.5 \, \text{N}$ and is at 31° to the horizontal.
(a) Resolve both vectors into components acting horizontally and vertically. **(4 marks)**
(b) Comment on the horizontal components of T_A and T_B. **(2 marks)**

A   B

Adding vectors

You can find the **resultant** of two vector quantities acting along the same line simply by adding them. For other cases, you must use must use scale drawing or calculation.

Forces with the same line of action

Two forces act along the same line of action

To distinguish the direction of the forces we choose a direction as positive, say to the right. The **resultant** or net force is then given by $(F_R - F_L)$. If the resultant is positive, the team on the right are pulling harder and the rope starts to move to the right If the answer is negative, the team on the left are making the rope start to move to the left.

Resultant by drawing

Coplanar vectors can be combined using scale drawings (Note that two vectors are always coplanar). You will need this method whenever a vector triangle is not right-angled.

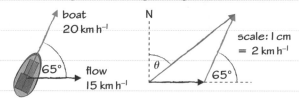

Accurately draw the two vectors to scale, end to end, and construct a third arrow (green), joining them. This represents the resultant.

Complete the triangle as shown to find the resultant velocity. The magnitude is found from the length, 29.6 km h^{-1}, and the bearing from north by measuring θ (N 52° E).

Worked example

A boat is motoring north with a velocity of 20 km h^{-1} across a river flowing east with velocity 15 km h^{-1}. Find the resultant velocity of the boat. **(2 marks)**

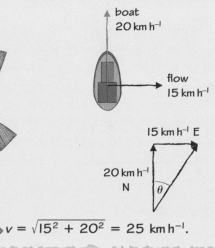

As these vector quantities are not collinear, acting along the same line, you cannot combine them by adding or subtracting their magnitudes. You must take their directions into account using a vector method.

When the vectors are at right angles to each other, we can use Pythagoras' theorem to find the resultant velocity.

The direction of the resultant velocity is found using trigonometry:
$v = \tan^{-1} 0.75 = \text{N } 36.9° \text{ E}$

$v = \sqrt{15^2 + 20^2} = 25 \text{ km h}^{-1}.$

Now try this

1 The diagram shows a simple pendulum at an instant during its oscillation. The pendulum bob weighs 8 N and the tension in the string, which is at 37° to the vertical, is 10 N.

Find the resultant force acting on the pendulum bob at the instant shown by drawing a vector triangle to scale. **(3 marks)**

2 A boat has a velocity of 5 m s^{-1} and sails on a bearing of N 25° E. The tide has a velocity of 2 m s^{-1} and a direction of N 100° E. Use a scale drawing to find the resultant velocity of the boat. **(3 marks)**

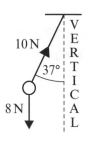

Projectiles

Anything that is launched into the air can be considered a **projectile**. Projectiles are subject to gravity and air resistance (the air exerts a force opposing motion, which varies with the direction of motion and the speed), but we often simplify the motion of projectiles by ignoring air resistance.

An example of a projectile

The SUVAT equations work for objects undergoing uniform acceleration. For a cannonball, the acceleration due to gravity is constant, and we can assume that air resistance is negligible

To use SUVAT equations for projectile motion:

• resolve the initial velocity into a vertical and a horizontal component: $v_V = u\sin\theta$ and $v_H = u\cos\theta$

• apply the gravitational force only to the vertical component of velocity – as it acts vertically downward, the horizontal component is not affected.

A cannonball is fired with an initial velocity of u m s^{-1} at an angle $\theta°$ to the horizontal. You can use the SUVAT equations to find how high, how far and for how long the cannonball flies.

A cannonball is fired at $150\,\text{m s}^{-1}$ at an angle of $30°$ to the horizontal.

(a) How high does it reach above the ground? (Acceleration due to gravity $g = 9.81\,\text{m s}^{-2}$.)

(3 marks)

To calculate how high (vertical displacement, s) the ball travels:

$v^2 = u^2 + 2as$ in which $a = -9.81\,\text{m s}^{-2}$

$0^2 = (75)^2 + 2 \times (-9.81)s$

$s = 287\,\text{m}$

(b) How far does it travel and how long is it in flight? **(2 marks)**

The time of flight will be twice the time t to reach the top of the trajectory.

$v = u + at$

$0 = 75 - 9.81t$

$t = 7.65\,\text{s}$

Range (horizontal distance travelled) = horizontal component of velocity × time of flight.

$\therefore$ range $= 150\cos 30° \times (2 \times 7.65)$

$= 1990\,\text{m}$

Deal with the motion in two parts. Here the first part of the question asks you to start with the vertical motion.

The initial vertical component of the cannon ball's velocity (u) is $150\sin 30° = 75\,\text{m s}^{-1}$. At the top of its flight the ball's vertical velocity (v) is zero for an instant before it starts to fall back to earth.

Choosing up as positive makes downward acceleration negative. Substitute values and solve.

How far the cannonball travels depends on how long it spends in flight, so you need to look at the vertical motion and work out how long the ball takes to come down. Then you can work out how far it travels horizontally in that time.

The horizontal component of velocity is constant, because without air resistance there is no horizontal force on the cannonball.

The answer should be given to the same number of significant figures as you were given in the question, so in this case no more than 3 significant figures.

A catapult launches a ball from the ground with a velocity of $200\,\text{m s}^{-1}$ at an angle of $75°$ to the horizontal. Find:

(a) the vertical and horizontal components of the initial velocity **(2 marks)**

(b) the time it is in flight before striking the ground again **(4 marks)**

(c) the range of the catapult ball (distance travelled horizontally). **(2 marks)**

Free body diagrams

Free body diagrams are used to visualise all the forces that act on a particular object.

Forces on a small body

A book on a non-horizontal rough surface may accelerate or not, depending on the balance of all the forces that act on it.

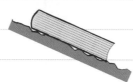

The word 'rough' indicates that friction will act to oppose motion. If we wish to consider zero friction surfaces we describe them as 'smooth'.

Free body diagrams simply show all the forces that act upon a small object. Each force is represented by a labelled arrow in its direction of action. The body is usually shown as a point.

Worked example

Draw a free body diagram for the book on the rough surface shown on the left.   **(3 marks)**

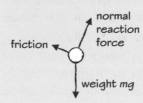

The arrows in this free body diagram do not represent vectors as the relative sizes of the forces are not known but the directions should be correct.

An extended but rigid body

When we study the effects of forces on small bodies or particles we usually think only about what happens to their state of motion in a straight line. However, when forces are applied to extended (long) bodies the result may make the body rotate about some point. The free body diagram must now show the body isolated from its surroundings with all the forces that act on the body in their correct positions relative to each other.

For the plank shown here, the weight will act downward through the **centre of gravity** of the plank. The friction force F_A acts upwards at A to oppose the tendency of the plank to slide down the wall, and similarly F_B acting on the plank at B opposes the tendency of the plank to slide away from the wall. N_A and N_B are the normal reaction forces acting on the plank at points A and B, respectively.

Again the actual sizes of these forces may not be known initially, so the free body diagram cannot represent magnitudes at this stage in the analysis of the problem.

The free body diagram for the plank

Remember to label each force. You may not be given the size of the forces but the directions should be correct. As the balloon is at rest the net or resultant force will be zero.

Now try this

1 Draw the free body diagram showing the forces that act on a helium balloon tied to the ground with a string.   **(3 marks)**

2 Draw the free body diagram for a parachutist in free fall (that is, parachute not yet opened).   **(2 marks)**

3 Draw the free body diagram for a brass sphere at rest in a horizontal v-shaped groove as shown. (In this question the angle of the notch is not specified.)   **(3 marks)**

Newton's first and second laws of motion

Newton described the effect that forces have on the state of motion of objects.

Newton's first law of motion

Objects remain at rest or continue to move in a straight line at constant speed unless they are acted on by a resultant force.

The first part of Newton's first law is easy to understand, but your common sense may tell you that moving objects tend to slow down and stop. In fact, this is because there **are** forces acting on them, like friction and air resistance.

Newton's second law of motion

$\Sigma F = ma$

ΣF is the vector sum of the forces acting on a body (Σ, the Greek capital letter sigma, means 'sum of'); m is mass, a is acceleration.

If the resultant force on an object is zero ($\Sigma F = 0$), then, as Newton's first law says, the object does not accelerate: $a = 0$.

This equation is easy to use as long as m is constant.

How force and mass affect acceleration

Consider two jet cars. Imagine negligible friction between the road surface and the wheels. The resultant force on the jet cars will therefore be the **thrust**, T_1, of the jet motor.

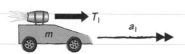
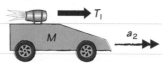

Here both jet cars have the same resultant force, T_1, acting to the right. The larger mass M of the jet car on the right means that it accelerates more slowly. Acceleration $a_1 >$ acceleration a_2.

The thrust on both jet cars is increased to T_2. This increases the rate of acceleration of both cars, so $a_3 > a_1$ and $a_4 > a_1$.

Terminal velocity

In the examples above you ignored forces opposing, (acting to the left of) the thrust of the motors. In a real situation friction in the wheel bearings and air resistance will oppose the motor thrust. The faster the jet car travels, the more air it has to displace per second, so the greater the opposing force of air resistance gets. Eventually the forces opposing the thrust will be as large as the thrust of the motor and the **resultant force** on the cars will be zero. From $\Sigma F = ma$, if $\Sigma F = 0$ then a must be zero too. The car will keep moving at constant velocity; it has reached its maximum or **terminal velocity**.

Now try this

1 An object with a mass of 600 g accelerates in a straight line at $5.0 \, \text{m s}^{-2}$. What resultant force must act on it to cause this acceleration? **(2 marks)**

2 A water skier with a mass of 50 kg experiences a resultant force of 125 N. Calculate her acceleration. **(2 marks)**

3 An aircraft has engines that can provide a maximum thrust of 180 000 N. It has a maximum acceleration of $1.8 \, \text{m s}^{-2}$.

(a) Calculate the maximum, fully loaded, mass of the aircraft. **(2 marks)**

(b) Explain why the aircraft will reach a maximum speed. **(3 marks)**

(c) Describe the effect on the maximum acceleration of having only half the maximum number of passengers on board.

'Describe' does not require a calculation or an explanation.

(1 mark)

Measuring the acceleration of free fall

Free fall means an object falling without experiencing air resistance.

Weight and free fall

The rate of acceleration of a falling object will depend on the local strength of the gravitational field. Gravitational field strength is defined as the force per kilogram that acts on an object:

$$g = \frac{F}{m}$$

On the Earth's surface g is $9.81 \, \text{N} \, \text{kg}^{-1}$ (though this varies a little from place to place).

This means that the **weight** of any object is given by

$$W = mg$$

Since the acceleration, a, of any object is given by $\frac{F}{m}$ and $F = W$, all objects fall with same acceleration, g, $9.81 \, \text{m} \, \text{s}^{-2}$, which is what is measured with the apparatus shown here.

> **Be careful!** Although the acceleration of freely falling objects and the gravitational field strength have same numerical value they are not the same thing!

The experiment

The electronic timer measures short time intervals with adequate accuracy. The steel ball completes the circuit at the top. When it is released the timer starts. The timer stops when the trapdoor at the bottom is knocked open, breaking a second circuit. The ball is dropped through a range of distances up to a metre, and the time for each distance recorded.

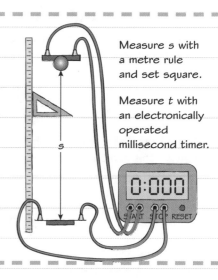

Measure s with a metre rule and set square.

Measure t with an electronically operated millisecond timer.

The graph of s against t^2

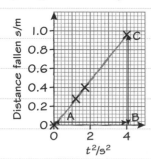

$s = ut + \frac{1}{2}at^2$, where $u = 0$ and $a = g$. Thus $s = \frac{1}{2}gt^2$, so plotting s against t^2 should produce a straight line graph with gradient $= \frac{g}{2}$.

The graph shown here has just a few points plotted and the line of best fit drawn. You should have **at least** 6 plotted points to produce a good straight-line graph.

To find a value for the acceleration of free fall from this graph, find the gradient by drawing a large triangle, ABC, and measuring AB and BC against the axis scales: AB = $0.97 \, \text{m}$ and BC = $0.2 \, \text{s}^2$.

This gives the gradient $\frac{\text{BC}}{\text{AB}} = \frac{0.97}{0.2} = 4.85 \, \text{m} \, \text{s}^{-2}$.

This is $\frac{1}{2} g$. Therefore this experiment yields a value for g of $9.7 \, \text{m} \, \text{s}^{-2}$.

Now try this

1. In mechanical bathroom scales, weight presses down on some stiff springs, and a pointer moves in proportion to how much the spring has compressed to give a reading calibrated in kilograms.
   (a) If the reading is 55 kg what is the actual weight? **(2 marks)**
   (b) If you took the same scales to the Moon, the scales would read 9.11 kg for the same mass.
   Explain with the aid of a calculation why this is so. **(3 marks)**

2. You have carried out the experiment described above. Explain how you would ensure the accuracy of the measurement of: (a) the distances fallen; (b) the time taken for each fall; (c) the value obtained for g. **(6 marks)**

Newton's third law of motion

Newton's third law: for every interaction between two objects there is an equal and opposite reaction.

Action and reaction forces

- Each force acts on a different body.
- Each force is of the same type.
- The magnitude of each force is the same.
- The forces are **collinear** – acting in the same line.
- The forces are opposite in direction.

Action and reaction

As a simple demonstration of the third law, imagine two people on skateboards. Amir actively pushes Bob and, as expected, Bob slides off to the right. However, Amir also moves, in the opposite direction. He has experienced a reaction force in the opposite direction to the action, the force he applied to Bob. If the two have the same mass, they move with the same initial speed showing that each force has the same magnitude.

Worked example

Ignoring air resistance and upthrust, describe the action–reaction pair for a ball in free-fall. **(2 marks)**

$F_1 = -F_2$

The ball experiences the gravitational pull of the Earth and the Earth experiences an upward pull due to the ball. The forces act on different bodies and are both gravitational forces.

Note that the ball is not in equilibrium; it accelerates downward because the only force acting on the ball acts downwards.

Worked example

A book is resting on a level table. Describe the action–reaction pair **(4 marks)**

pull of book acting on Earth | normal contact force N_t of table on book

pull of Earth acting on book (weight) | normal contact force N_b of book on table

There are two action–reaction pairs: weight and contact force.

Weight: Gravity pulls down on the book. The reaction to this is the upward pull of the book on the Earth.

Contact force: The table pushes upwards on the book. The reaction to this is the downward contact force of the book on the table. In this case the book is in equilibrium and at rest.

Now try this

1 Two balloons have been electrically charged by rubbing with a dry cloth. They are both suspended by a light nylon string from the same point and are shown at rest in the diagram to the right.
(a) Describe each force that acts on the left-hand balloon. **(3 marks)**
(b) For each force explain the nature and location of the reaction force. **(3 marks)**

2 A parachutist jumps out of a plane and accelerates until reaching terminal velocity. When she opens her parachute there will be a period of deceleration before she reaches a new, slower terminal velocity. Describe the action–reaction pairs on the parachutist during her descent. **(3 marks)**

13

Momentum

Momentum is a vector property of moving objects.

Mass and velocity

Momentum, p, is the product of the mass, m, of an object and its velocity, v: $p = mv$

Newton's second law can be stated in terms of momentum:

The **rate of increase of momentum** of a body is proportional to the resultant force that acts on the body and takes place in the direction of the resultant force.

If you use S.I. units, the constant of proportionality is 1.

Look back to page 11 to read about Newton's second law expressed in the form $F = ma$.

Derivation of $F = ma$

A body of mass m travelling with a velocity u has a resultant force F acting on it for t seconds. After this period of acceleration the body travels at velocity v.

If mass m is constant, the increase in momentum of the body is: $mv - mu = m(v - u)$

so the rate of increase of momentum is $\dfrac{m(v - u)}{t}$

so $F = \dfrac{m(v - u)}{t}$

or $F = ma$ because $\dfrac{(v - u)}{t}$ is acceleration

Newton's law of motion

Expressed in terms of momentum:

1 Newton's first law: the momentum of a body remains constant unless a net force acts on the body

2 Newton's second law: a change of momentum is proportional to the applied force and takes place in the same direction

3 Newton's third law: colliding bodies exert equal and opposite forces on each other, so total momentum is unchanged.

Conservation of momentum

force exerted on A by B ← → force exerted on B by A

Consider two objects travelling towards one another along the same straight line. When they collide, each exerts a force on the other while they are in contact. These two forces are an action–reaction pair, equal in magnitude and opposite in direction and acting on different objects.

The equation $F = ma$ can be rearranged to $Ft = mv - mu$; in other words, the **change in momentum** of a body is determined by the size of Ft.

In any collision, $F \times t$ for the two bodies is the same magnitude but opposite in direction, because the forces on each body act in opposite directions. The change in momentum Ft is a vector quantity.

So increase in momentum of B = the decrease in momentum of A. This is the **law of conservation of momentum**: In any collision the total momentum of the colliding bodies remains constant, provided no external forces act on the bodies.

Worked example

A block of ice of mass 1.0 kg slides across a frozen pond at 6.0 m s⁻¹ and collides with a stationary block of ice of mass 2.0 kg. After the collision the 2.0 kg block moves off with a velocity of 4.0 m s⁻¹ in the same direction. Assuming that friction is negligible, calculate the velocity of the 1.0 kg block after the collision. **(3 marks)**

Total momentum before collision
= total momentum after collision

$(1.0 \times 6.0) + (2.0 \times 0) = (1.0 \times v) + (2.0 \times 4.0)$

$v = (6 - 8) = -2.0 \text{ m s}^{-1}$

The minus sign shows that the 1 kg block rebounds, moving in the opposite direction to its initial motion.

Now try this

1 In a game, a block of ice, A, with a mass of 4.0 kg is sliding at 1.0 m s⁻¹ over the surface of an ice rink. Another block, B, of mass 2.0 kg, is kicked in the same direction at 3.0 m s⁻¹, in order to hit block A. After the collision A is travelling in the same direction at 2.0 m s⁻¹. What is the velocity of B after the collision? **(4 marks)**

2 A railway wagon of mass 200 kg is shunted along a track at 6.5 m s⁻¹. It meets a second, stationary, wagon and couples up with it. The coupled wagons move together at 2.6 m s⁻¹ in the same direction. What is the mass of the second wagon? **(4 marks)**

When two objects collide and stick together the collision is termed inelastic. Momentum is still conserved, provided no external forces act.

Moment of a force

The moment of a force is a measure of its turning effect.

The principle of moments

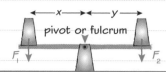

A see-saw will be balanced, or in **equilibrium**, if
$F_1 = F_2$ and the distances x and y are equal.
If either the weight F_2 or the distance y is
decreased (or both) the see-saw will rotate
anticlockwise about the pivot.

The turning effect or **moment of force** $= Fx$
where x is the **perpendicular distance** from the
pivot to the **line of action of the force**.

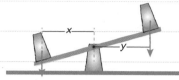

> Be sure to use the perpendicular distance
> from the pivot to the line of action, as shown.

The **principle of moments** states that a body will
be in rotational equilibrium if the sum of clockwise
moments acting on it is equal to the sum of the
anticlockwise moments, provided that moments
are taken about the same point.

Worked example

The system shown here is balanced. Find the size of
the force labelled F. **(3 marks)**

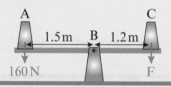

For equilibrium, clockwise moment =
anticlockwise moment (about any point).

Taking moments about B:

$F \times 1.2\,\text{m} = 160\,\text{N} \times 1.5\,\text{m}$

$F = \dfrac{160 \times 1.5}{1.2} = 200\,\text{N}$

'Hidden' forces

The principle of moments works no matter what
point you choose to take moments about. In the
worked example, taking moments about C:

↺ moment $= 160 \times (1.5 + 1.2)$, so how is
equilibrium possible? The answer is that there is
an upward reaction force R on the see-saw at B
providing a ↻ moment, so
$$R \times 1.2 = 160 \times (1.5 + 1.2) = 360\,\text{N}.$$

Note that 360 N equals the sum of the two
downward forces on the see-saw – equilibrium also
requires that there is no **resultant** force acting.

We have assumed here that the see-saw itself is
weightless. Generally this will not be so!

Centre of gravity

With care it is possible to balance a metre rule by
supporting it at the middle. In equilibrium there
must be no net force on the rule in any direction
and no net turning moment. Therefore, the weight
must act downwards through a point in the middle
of the rule.

This point is called the **centre of gravity** of the
body, the point through which all of the weight of
a body appears to act. For bodies with regular
shape made of uniform material the centre of
gravity is at the geometric centre.

Now try this

A diving board has a weight of 200 N and is 2.0 m
long. It is supported by two steel supports placed at
the end A and at B, 0.80 m from A. It is tested with a
500 N weight at the other end.

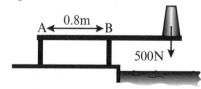

(a) Draw a free body diagram for this
system. **(2 marks)**

(b) Take moments about B to determine the
magnitude and direction of the force due to the
support at A. **(4 marks)**

(c) Take moments about A to determine the
magnitude and direction of the force due to the
support at B. **(4 marks)**

Exam skills 1

This exam-style question uses knowledge and skills you have already revised. Have a look at pages 7, 8, 13 and 15 for a reminder about vectors, forces, moments and equilibrium.

Worked example

The diagram below shows a picture hanging by two strings from a hook on a wall. The tension in each string is T.

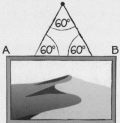

A   60° 60°   B

The mass of the picture is 920 g.

(a) Explain why the tension in both strings must be equal. **(2 marks)**

The picture must be in equilibrium, so the horizontal forces exerted on the picture by each string must be equal and opposite. The horizontal component of tension from one string is $T\cos 60°$. They are both at the same angle so must have the same tension.

(b) Calculate the tension in one of the strings. **(3 marks)**

The picture is in equilibrium so the total upward force ($2T$, twice the tension in each string) must balance the weight of the picture:

$2T\sin 60° = mg$

$T = \dfrac{mg}{2\sin 60°} = \dfrac{0.920 \times 9.81}{2\sin 60°} = 5.21\,\text{N}$

(c) State and explain what would happen to the tension in the strings if they were connected to the corners A and B of the picture rather than the positions shown. **(3 marks)**

The tension would increase. $T = \dfrac{mg}{2\sin \theta}$, where θ is the angle to the horizontal. When the strings are moved to the corners of the picture, θ gets smaller so T must increase.

(d) Explain why the centre of mass of the picture must lie vertically below the supporting nail. **(3 marks)**

When the centre of mass is directly below the nail the line of action of the weight acts through the nail. The lines of action of the forces in both strings also act through the nail. If all three forces acting on the picture pass through the same point, then there is no resultant moment, no turning effect, so the moments are in equilibrium.

Command words: 'state and explain'

Read the question carefully – when you are asked to state **and** explain, there will usually be marks for **both** parts of your answer.

Remember that for an object to be in equilibrium both horizontal **and** vertical forces must add to zero.

Maths skills Make sure you show each part of your working when carrying out a calculation. Usually this means you must:
- ✓ State the equation.
- ✓ Rearrange the equation.
- ✓ Substitute values from the question.
- ✓ Calculate an answer.

Take care to convert non-S.I. units to S.I. units before carrying out your calculations. In this case the mass is given in grams; you must convert to kilograms in order to get a weight in newtons.

Maths skills If you introduce a new variable, such as θ, it is important to state what it represents. Here, it is the angle of the string to the horizontal.

You might be tempted to answer this part by simply saying that this is necessary for the picture to be in equilibrium. A better answer would be to explain that it is so that there is no resultant moment acting on the picture – the moments are in equilibrium.

There are often alternative ways to answer a question. In this case, instead of describing the equilibrium condition, you could explain that if the centre of gravity was not vertically beneath the nail then there would be a resultant moment acting on the picture so it would not be in equilibrium.

Work

In physics, 'work' has a very specific meaning.

The definition of work

$$\Delta W = F\Delta s$$

ΔW ('delta W') means the increase in work done by a force F when it is applied through a distance Δs in the direction of the applied force.

Work is a scalar quantity and is measured in joules (J, equivalent to N m).

Resolving forces

Sometimes the force applied to move an object is not acting along the **line of motion** (the direction of movement) of the object to which the force is applied.

See page 7 to review how to resolve vectors into two perpendicular components.

Here the pulling force F has a horizontal component acting along the line of motion of the sledge and a vertical component acting in the direction in which there is no movement.

The component of the force acting in the line of movement is $F\cos\theta$, so the work done by the force in moving a horizontal distance Δs is:

$$\Delta W = F\cos\theta \times \Delta s$$

Worked example

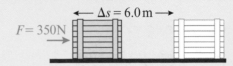

Calculate how much work is done when a force of 350 N is applied to move a crate 6.0 m across a rough horizontal surface. **(2 marks)**

$\Delta W = F\Delta s$

$\Delta W = 350 \times 6.0 = 2100\,J$

Worked example

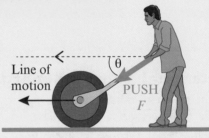

A gardener pushes a garden roller with a force of 150 N at 60° to the horizontal.

How much work is done if the roller is pushed 110 m along the lawn? **(2 marks)**

Horizontal component of

$F = 150\cos 60° = 75\,N$

Distance moved horizontally = 110 m

$\therefore \Delta W = 75 \times 110 = 8250\,J$

Worked example

A boy pulls a sledge along with a rope. The force of tension in the rope is 80 N acting at 45° to the horizontal.

The boy pulls the sledge for 0.50 km. Calculate the work he has done. **(3 marks)**

Horizontal component of

$F = 80\cos 45° = 57\,N$

Distance moved horizontally = 500 m

$\therefore \Delta W = 57 \times 500 = 28\,500\,J$

Now try this

1. A lift with a gross load of 6000 N travels 200 m to the top of an office block at constant speed. Calculate the work done on the lift. **(2 marks)**

2. A passenger aircraft cruises at a velocity of 800 km h^{-1} in level flight for 3.0 hours. Its engines provide a thrust of 700 000 N. How much work is done by the plane during this flight? **(4 marks)**

3. A block of wood weighing 5.0 N slides 150 cm down a rough plane inclined at 20° at a constant speed. Calculate the work done by the block. **(3 marks)**

weight 5N 20°

Kinetic energy and gravitational potential energy

Gravitational potential energy (E_{grav}) is the energy an object has by virtue of its position in a gravitational field. Kinetic energy (E_k) is the energy an object has by virtue of its movement.

Kinetic energy

If a body of mass m is accelerated from rest ($u = 0$) by a force F then its acceleration $a = \frac{F}{m}$. This is Newton's second law.

From $v^2 = u^2 + 2as$ we see that $v^2 = 2as$, so

$$v^2 = \frac{2Fs}{m}$$

which, when rearranged, gives

$Fs = \frac{1}{2}mv^2$. Fs (= ΔW) is the increase in the energy of the body due to the work done on it, hence

$$E_k = \frac{1}{2}mv^2$$

E_k, like work, is a scalar quantity measured in joules.

Worked example

A bullet with a mass of 20 g is fired from a rifle with a barrel 80 cm long with a velocity of $500\,\text{m s}^{-1}$.

(a) What is the kinetic energy of the bullet? **(2 marks)**

$E_k = \frac{1}{2}mv^2$

$E_k = \frac{1}{2} \times 0.020 \times (500)^2 = 2500\,\text{J}$

(b) What is the average force on the bullet whilst it is accelerating along the barrel? **(2 marks)**

$2500\,\text{J} = \Delta W$, the work done on the bullet by the average force in the barrel

average force $F = \dfrac{\Delta W}{s} = \dfrac{2500}{0.8} = 3100\,\text{N}$

Gravitational potential energy

An object of mass m has a weight mg. At the Earth's surface $g = 9.81\,\text{N kg}^{-1}$.

Lifting this object through a distance h requires work to be done on the object. $\Delta W = Fs$, so here $\Delta W = (mg) \times h$

The work done on the mass has increased its gravitational potential energy.

$$\Delta E_{grav} = mgh$$

This energy is stored or potential energy that can be transferred by releasing the mass.

Worked example

A student of mass 50 kg climbs 25 steps up a tall ladder. The rungs on the ladder are 30 cm apart. What is the increase in the student's gravitational potential energy when at the top of the ladder? **(2 marks)**

Total height climbed = $0.30 \times 25 = 7.5\,\text{m}$

$\Delta E_{grav} = mgh$

$\Delta E_{grav} = 50 \times 9.81^{-1} \times 7.5 = 3700\,\text{J}$

Practical applications of energy transfer

Hydroelectric power stations make use of gravitational potential energy stored in the large mass of water collected in high reservoirs in mountainous regions. The water can be allowed to fall, under gravity, gathering speed and thus kinetic energy as it falls. This kinetic energy is then transferred by turbine generators into electrical energy.

Longcase 'grandfather' clocks are powered by the transfer of gravitational potential energy stored in heavy weights that have been raised to the top of the case. As they fall, controlled by the clock mechanism, the weights transfer the stored energy into the movement of the clock by turning a wheel.

Now try this

1 A block of wood of mass 1.2 kg slides 150 cm down a smooth plane inclined at 37° at a constant speed.

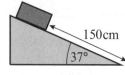

(a) Calculate how much gravitational potential energy has been transferred when the block reaches the bottom of the slope.

(b) Assuming all the gravitational potential energy has been converted to kinetic energy when the block reaches the bottom of the slope, calculate how fast the block is moving. **(4 marks)**

2 Calculate the kinetic energy of a car with a mass of 1200 kg travelling at $108\,\text{km h}^{-1}$. **(3 marks)**

3 A firework is launched vertically with a velocity of $40\,\text{m s}^{-1}$. Calculate how high it will travel, stating the assumptions you have made. **(3 marks)**

Conservation of energy

The energy of a closed system remains constant.

The principle of conservation of energy

In a closed system this is one which nothing can enter or leave the total energy remains constant, although it can be transferred within the system.

A motor and pulley system lifting a load is an example of an energy conversion process:

- electrical energy is converted to kinetic energy by a motor
- the energy is transferred to the load, increasing its kinetic and gravitational potential energy.

Energy is converted from one form to another, but the total amount of energy supplied in electrical form is fully accounted for; none is 'lost', though some may be converted to forms that you don't want, such as heat and noise. This is the **principle of conservation of energy**.

Worked example

An object of mass 2.0 kg is raised to a height of 30 m above the ground and then dropped.

(a) Describe the energy changes that take place from the moment the object is released until after it has come to rest on the ground. **(4 marks)**

The object has gained gravitational potential energy (GPE) $E_{grav} = mg\Delta h$
$(= 2.0 \times 9.81 \times 30\,J)$ from being raised.

As it falls, its GPE decreases (h decreases) and it gains kinetic energy (KE) as it accelerates.

At any given moment throughout the fall, by the principle of conservation of energy, loss of GPE = gain in KE.

On impact all the GPE the object gained when it was raised has been transferred into KE. It is assumed there is no air resistance.

During the impact the energy is converted into sound, heat and deformation of the ground.

(b) Use the principle of conservation of energy to calculate the speed with which it hits the ground. **(3 marks)**

GPE lost = KE gained
$mg\Delta h = \frac{1}{2}mv^2 \rightarrow v = \sqrt{2g\Delta h}$

| Note that mass cancels. |

$\therefore v = \sqrt{(2 \times 9.81 \times 30)} = 24.3\,m\,s^{-1}$

Worked example

The diagram shows a railway truck hitting a buffer. The buffer spring is compressed by 12.5 cm when the truck is brought to rest.

$v = 0.50\,m\,s^{-1}$

12.5 cm

$m = 2.5 \times 10^3\,kg$

(a) What is the kinetic energy of the moving truck? **(3 marks)**

$E_k = \frac{1}{2}mv^2 = \frac{1}{2} \times 2.5 \times 10^3 \times (0.50)^2$
$= 312.5 = 310\,J$ to 2 s.f.

(b) What is the average force F exerted by the buffer, assuming all the truck's kinetic energy is converted to stored energy in the buffer spring? **(3 marks)**

Work compressing the spring = F × distance

$F \times 0.125 = 312.5$
$F = 2500\,N$

Make sure you convert to S.I. units where necessary.

In practice energy would be converted to other forms as the truck was halted.

Now try this

A stone of mass 50 g is thrown with a velocity of 50 m s⁻¹ at an angle of 53° to the horizontal. Assume air resistance is negligible and use $g = 9.81\,N\,kg^{-1}$.

(a) Calculate how much work must be done on the stone. **(2 marks)**

(b) State the type(s) of energy the stone has at the top of its flight. **(2 marks)**

(c) Calculate how high the stone will travel. **(3 marks)**

Work and power

Like work, power has a strictly defined meaning in physics: the rate at which work is done.

Work and energy

In physics **work** is defined as:

work done = force × distance moved

$$\Delta W = F\Delta s$$

Two men push crates the same distance across the floor at the same steady speed. The bigger, heavier crate needs more force, because of greater frictional resistance. The man pushing the larger crate does more work. In other words, he transfers more energy.

Power

The man pushing the larger crate has done more work than the man pushing the small crate in the same time. He has a greater **power** output.

Power can be defined in two ways:

 The rate of energy transfer:

$$P = \frac{E}{t}$$

 The rate of doing work:

$$P = \frac{W}{t}$$

Power is measured in watts (W).

$$1\,W = 1\,J\,s^{-1}$$

Worked example

1 A man pushes a box at a steady rate of $2.5\,m\,s^{-1}$ for 12 seconds by applying a force of 80 N. Calculate the work he does and his power output. **(3 marks)**

The distance through which the push of 80 N is applied is $2.5\,m\,s^{-1} \times 12\,s = 30\,m$

$\therefore \Delta W = 80 \times 30 = 2400\,J$

$$P = \frac{W}{t} = \frac{2400}{12} = 200\,W$$

2 A forklift truck lifts a 250 kg pallet and load through 180 cm in 1.2 s. Calculate the work done and the power of the forklift. **(3 marks)**

$W = F\Delta s = mg\Delta s$

$= 250 \times 9.81 \times 1.8 = 4414.5$
or 4400 J to 2 s.f.

$$P = \frac{W}{t} = \frac{4414.5}{1.2} = 3700\,W \text{ or } 3.7\,kW$$

> Remember to convert non-S.I. units like centimetres into S.I. units.

> This is also the energy transferred by the forklift.

> Here 'power of the motor' means the rate of conversion of electrical energy to mechanical energy.

3 An electric motor raises a 600 kg lift at $3.0\,m\,s^{-1}$. Assuming no energy is wasted, calculate the power of the electric motor. **(3 marks)**

In one second the motor does work = $mg\Delta h$
$= 600 \times 9.81 \times 3 = 17658\,J$.

Motor power = 17.7 kW.

> Power is the rate of energy transfer: that is, energy transferred per second, $P = \frac{E}{t}$.
> In one second the lift is raised 3 m and gains gravitational potential energy $mg\Delta h$.

Now try this

A student of mass 45 kg runs up a flight of 30 steps in 15 s. Each step is 20 cm high. $g = 9.81\,N\,kg^{-1}$.

(a) What is the student's increase in gravitational potential energy? **(2 marks)**

(b) What is the student's power output in watts? **(2 marks)**

Efficiency

The efficiency of an energy transfer is the fraction of energy supplied that is transferred usefully.

Useful energy

When you use an electric kettle to boil water, the **useful energy** is that which raises the temperature of the water; the **wasted energy** is that which escapes from the water, heating up the kettle, its element and the surroundings of the kettle, and in the form of noise. Well-designed, **efficient** kettles reduce the amount of energy that is not doing what we want it to do – heating the water.

Efficiency

$$\text{Efficiency} = \frac{\text{useful energy output}}{\text{total energy input}}$$

which is the same as

$$\text{efficiency} = \frac{\text{useful power output}}{\text{total power input}}$$

Multiply by 100 to get efficiency as a percentage.

Clearly efficiency cannot exceed 1 (one), or 100%.

Worked example

This kettle uses 360 000 J of electrical energy to heat some water to boiling point.
The energy actually transferred into the water to bring it to the boil is 300 000 J.
What is the efficiency of the kettle?   **(4 marks)**

$$\text{Efficiency} = \frac{\text{useful energy output}}{\text{total energy input}}$$

$$\text{Efficiency} = \frac{300\,000\,\text{J}}{360\,000\,\text{J}} = 0.83 \text{ (or 83\%)}$$

Worked example

Lamp A is a tungsten filament lamp. These are only 5% efficient. Lamp B is a compact fluorescent lamp. These are claimed to use 75% less energy than filament bulbs. Lamp A is rated at 60 W.

Lamps A and B are both in use for 2.0 hours.

(a) Find the total electrical energy input to lamp A in joules. Calculate the useful light output of lamp A in joules and say how the remaining amount is 'wasted'.   **(4 marks)**

$E = P \times t$

Energy input = $60 \times (2 \times 3600)$

= 432 000 J

Useful output as light = 432 000 × 0.05

= 21 600 J

The remaining 410 400 J are converted into heat energy.

(b) Assuming that both lamps have the same useful light output and the maker's claim for B is accurate, calculate how much electrical energy lamp B uses in 2.0 h, and the efficiency of lamp B.   **(4 marks)**

If lamp B uses only 25% of the energy used by A:

Energy input = 432 000 × 0.25

= 108 000 J

$$\text{Efficiency} = \frac{\text{useful energy output}}{\text{total energy input}}$$

$$\text{Efficiency} = \frac{21\,600}{108\,000} = 0.20 \text{ (or 20\%)}$$

Now try this

1 An electric motor in a hoist is rated at 3.0 kW. It can lift a load of 1200 N at a speed of 1.5 m s⁻¹. Calculate the efficiency of this motor.   **(3 marks)**

2 A car is driving at a 60 mph (27 m s⁻¹) along a straight clear road. The forces opposing the motion of the car at this speed (air and rolling resistance) are, in total, 400 N. The car engine is approximately 20% efficient.
   (a) Calculate the useful power output of the car at this speed.   **(2 marks)**
   (b) Calculate the total power input to the car engine to maintain this speed.   **(2 marks)**
   (c) What is happening to the wasted power?   **(2 marks)**

Exam skills 2

This exam-style question uses knowledge and skills you have already revised. Have a look at pages 10–13, 18 and 19 for reminders about forces, energy and motion.

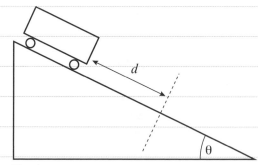
Fig. 1

Worked example

A student wants to investigate the transfer of gravitational potential energy to kinetic energy as a toy car moves down a slope. She does this by releasing the car from rest and measuring the speed v when it has moved a distance d along the slope as shown in Fig. 1. She records the speed of the car for a range of values of d and then plots a graph of v^2 against d.

(a) Describe an experimental method to measure the speed of the car when it has moved a distance d.   **(3 marks)**

Attach a card of measured length (e.g. 5.0 cm) to the top of the car and place a light gate at distance d. When the card cuts the light beam a data logger can use the time and length of card to calculate and display the speed.

Note the command word 'describe': this means that you should not just write 'use a light gate'. Explain **how** it is used.

Questions like this do not have one set answer. You could use a motion sensor or video capture and analysis software instead of a light gate, but whatever you choose to use you must describe **how** the apparatus is set up and used, and the data that it measures.

(b) Show that a graph of v^2 against d should be a straight line through the origin, if work done against friction can be neglected.   **(4 marks)**

Loss of GPE = gain of KE

$mgh = \frac{1}{2}mv^2$, where h = the height of the ramp, g = the acceleration due to gravity, m = the mass of the car, and v = its measured velocity.

$mgd\sin\theta = \frac{1}{2}mv^2$

$\qquad v^2 = (2g\sin\theta)\,d$

All terms inside the bracket are constant, so v^2 is directly proportional to d.

If the two variables plotted on the graph are directly proportional, like v^2 and d, the graph will be a straight line through the origin.

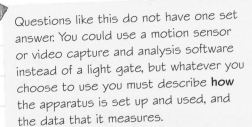

Maths skills To answer part (b) you need to:

1. Start with conservation of energy.

2. Write down the equation and define the terms.

3. Rearrange the equation.

4. Show that it leads to $v^2 \propto d$.

(c) Here is some data from the student's experiment:
mass of car = 0.40 kg
angle of slope to horizontal = 20°
distance d = 45 cm
speed v = 1.65 m s^{-1}
Use this data to calculate the average frictional force F acting on the car.   **(4 marks)**

The total input energy is from GPE. Since energy is conserved, any energy that is not transferred to KE must be transferred to heat by work done against friction.

Work done against friction = loss of GPE − gain in KE
= $mgd\sin\theta - \frac{1}{2}mv^2$
= 0.4 × 9.81 × 0.45 × sin 20° − 0.4 × (1.65)²/2
= 0.604 − 0.545 = 0.059 J

Work done against friction = Fd

F = 0.059/0.45 = 0.13 N

Start by stating the underlying physics – in this case it is the law of conservation of energy. Then use it to work out the total work done against friction. Finally use this intermediate result and the equation $W = Fd$ to calculate a value for F.

Basic electrical quantities

Values of current and potential difference are important in determining the behaviour of an electric circuit.

Current

When electrically charged particles move through a conductive material we refer to a **current**.
In metals the charged particles are electrons.

Electric current, I, is the rate of flow of charge, Q.

$$I = \frac{\Delta Q}{\Delta t}$$

The unit of current, the amp (A), is a rate of flow of charge of 1 coulomb (C) per second. $1\,A = 1\,Cs^{-1}$.

Potential difference (p.d.)

Potential difference is a measure of the work done per unit charge passing through a conducting element in a circuit.

$$V = \frac{W}{Q}$$

The unit of p.d. is the volt (V). For a p.d. of 1 V between two points in a circuit, 1 J of energy is transferred per 1 C of charge flowing between the points. $1\,V = 1\,JC^{-1}$.

Potential difference or p.d. is sometimes referred to as voltage. It is the amount of energy that a component transfers per unit of charge passing through it. A voltmeter it measures the difference in electric potential between two points in a circuit.

Worked example

A lamp has a current of 50 mA through it. Calculate the electric charge that passes through it in 1 minute. **(1 mark)**

Convert to S.I. units: 0.050 A and 60 s

$\Delta Q = I \times \Delta t$

$\Delta Q = 0.050 \times 60 = 3.0\,C$

Electrical power

Power is the rate of doing work, or the rate of energy transfer. You can express this definition in electrical terms using the equations above to find an equation for power in electric circuits:

$$P = \frac{W}{t} = \frac{QV}{(Q/I)} \text{ as } Q = I \times I$$

$$P = VI$$

S.I. units are watts (W): $1\,W = 1\,V \times 1\,A$

Since $P = \frac{W}{t}$, it follows that $W = Pt$

gives the amount of electrical energy transferred. See page 20 for a reminder of the relationship between power and work.

Worked example

The lamp in the example on the left was connected to a 6 V supply:

6 V ──→ 0.05 A ─(○)─ 0 V

How much energy is transferred into heat and light in the lamp if the lamp is on for 1 minute? **(2 marks)**

$W = V \times Q$

$W = 6.0 \times 3.0 = 18\,J$

See page 20 for the relationship between power and work.

Worked example

An electric heater operates from a 230 V supply and draws a current of 12.5 A.

(a) Calculate the power of this heater. **(1 mark)**

$P = VI = 230 \times 12.5 = 2875$
$\qquad\qquad\qquad\qquad\; = 2880\,W\ (3\ \text{s.f.})$

(b) Calculate how much energy is transferred into heat in 1 hour and 40 minutes by the heater. **(2 marks)**

Energy transferred $W = Pt$

$= 2875 \times (100 \times 60) = 17.3\,MJ\ (3\ \text{s.f.})$

Now try this

1　A reading lamp operating from a 230 V supply has a power of 60 W.

  (a) Calculate the current that it draws. **(1 mark)**

  (b) Calculate how much charge passes through it in 5.0 minutes. **(2 marks)**

2　A current of 250 mA passes through a torch lamp. After 30 minutes 2400 J of electrical energy has been transferred into light and heat.

  (a) Calculate the power of the lamp. **(2 marks)**

  (b) Calculate how much power is transferred in the lamp. **(2 marks)**

  (c) Calculate the voltage across the lamp. **(2 marks)**

Ohm's law

You will recall the equation $V = I \times R$ from your earlier studies. Ohm's law is a special case in which $I \propto V$ at constant temperature.

Ohm's law experiment and results

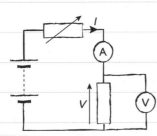

Circuit to demonstrate the relationship between p.d. across and current in a component

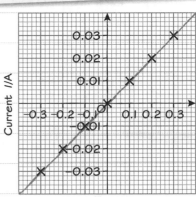

The results for an ohmic conductor

For a metallic conductor at constant temperature the graph of current against p.d. is a straight line passing through the origin. The resistance R, measured in ohms (Ω), can be calculated at any point by reading off values for I and V, and will be found to be constant provided that the temperature is unchanged.

Ohm's law: at constant temperature, the current through the conductor is proportional to the p.d. applied across it.

Non-ohmic components

Not all circuit components follow Ohm's law. You will be required to recognise, sketch and interpret graphs for some components that do not. These are called **non-ohmic components**.

As the temperature of the filament lamp increases, its resistance increases.

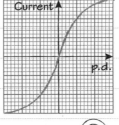

Filament lamp ─◯─

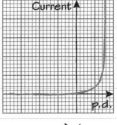

Diode ─▷┤─

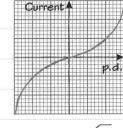

Thermistor ─⊏─

The **diode** has low resistance when 'forward biased', >0.6 V, whereas when $V < 0.6$ V its resistance is so high that it is virtually non-conducting.

The **thermistor** shown, which is negative temperature coefficient or NTC type, has a resistance that decreases as it gets hotter. A use for thermistors is covered on page 31.

Worked example

1 Find the current in a 12 kΩ resistor when 9.0 V is applied across it. **(1 mark)**

$$I = \frac{V}{R} = \frac{9.0}{1.2 \times 10^4} = 0.75 \text{ mA}$$

2 Find the resistance needed to allow a current of 200 μA when the resistor is connected across a 5.0 V supply. **(1 mark)**

$$R = \frac{V}{I} = \frac{5}{2 \times 10^{-4}} = 25 \Omega$$

Now try this

The current through a filament lamp is 10 mA when the p.d. across it is 0.10 V. When operated at its working p.d. of 6.0 V it draws 60 mA.

(a) Calculate the resistance of the lamp at these p.d.s. **(2 marks)**

(b) Explain why this change occurs. **(2 marks)**

Conservation laws in electrical circuits

The rules of conservation of charge and energy show how currents and p.d.s behave in series and parallel circuits.

Current in = current out

The current into a component, like a lamp or a resistor, must equal the current out of it, so $I_2 = I_4$ and $I_3 = I_5$. If this were not the case, charge would be disappearing or being created within the components.

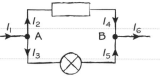

Similarly the current flowing toward and away from the nodes A and B must also balance. Therefore $I_1 = I_2 + I_3$, and $I_4 + I_5 = I_6$.

This is the consequence of **conservation of charge** in electrical circuits.

P.d.s add up in series and are the same across parallel branches

V is the amount of energy transferred per coulomb passing from A to B. V_1 is the amount of energy transferred per coulomb in R_1 and V_2 is the amount of energy transferred per coulomb in R_2. Each coulomb that passes from A to B passes through R_1 then R_2.

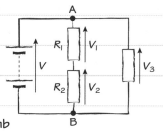

Energy conservation requires that the total amount of energy transferred as charge is moved from A to B must be equal to the energy transferred in R_1 and R_2. Therefore $V = V_1 + V_2$.

This is the consequence of **conservation of energy** in electrical circuits.

Since the potential difference between A and B must be the same through both possible paths, V_3 must be equal to V.

Now try this

1 Complete this sentence about electric current in circuits. **(3 marks)**

The total current out of any point in a circuit must .... the .... current into that point as .... must be conserved.

2 Complete this sentence about p.d.s in electric circuits. **(4 marks)**

The p.d.s across components in series must .... to the supply e.m.f. The p.d.s across components in parallel must be ....
The total .... supplied by a battery per coulomb of charge circulated must be .... the total energy transferred in the circuit.

Worked example

1 Calculate the current through the two resistors. Hence state the current, I, supplied by the battery. **(3 marks)**

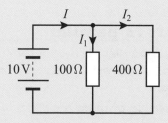

The p.d. across both resistors is 10 V.

$I_1 = \dfrac{10}{100} = 0.1\,A$; $I_2 = \dfrac{10}{400} = 0.025\,A$.

$I = I_1 + I_2 = 0.125\,A$

P.d.s across parallel branches of a circuit are the same.

2 Write equations for the power dissipated in a resistor in terms of (a) p.d. and current, (b) current and resistance, (c) p.d. and resistance. **(3 marks)**

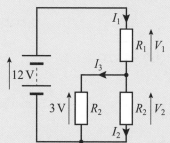

Know $P = VI$ and be able to derive $P = I^2R$ and $P = \dfrac{V^2}{R}$ from it.

(a) $P = VI$
(b) Since $P = VI$ and $V = IR$, $P = I^2 \times R$
(c) Since $P = VI$ and $I = V/R$, $P = V^2/R$

3 (a) State the values of V_1 and V_2 for the circuit shown above. **(2 marks)**

$V_2 = 3\,V$ (p.d.s between the same two points in a circuit must be equal), $V_1 = 9.0\,V$
($V_1 = 12\,V - V_2$).

   (b) Calculate the value of I_2 if $R_2 = 150\,\Omega$. **(2 marks)**

$I_2 = V_2/R_2 = 3/150 = 0.02\,A$

   (c) If $I_1 = 50\,mA$ state the value of I_3, and calculate the value of R_1. **(3 marks)**

$I_1 = I_2 + I_3$
$\rightarrow I_3 = I_1 - I_2$; $I_1 = 0.05\,A$ so $I_3 = 0.030\,A$
$R_1 = V_1/I_1 = 9/0.05 = 180\,\Omega$

Resistors

There are formulae to calculate the equivalent resistance of resistor combinations.

Resistors in series

The total p.d. across resistors in series will be the sum of their individual p.d.s. Take an example of three resistors R_1, R_2 and R_3 in series. The current I through each resistor must be the same. The total p.d. V is the sum of the p.d. across each resistor, $V_{total} = V_1 + V_2 + V_3$.

Finally, $V = IR_{total}$. Therefore: $IR_{total} = IR_1 + IR_2 + IR_3$

so $R_{total} = R_1 + R_2 + R_3$

Therefore, for resistors in series the total resistance is the sum of the individual resistances.

Resistors in parallel

Resistors in parallel will have a total current through them that is the sum of their individual currents. The p.d. V across each must be the same for each branch.

Take an example of three resistors in parallel, $I_{total} = I_1 + I_2 + I_3$

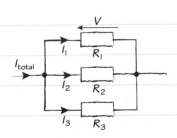

and $I = \dfrac{V}{R}$. Therefore: $\dfrac{V}{R_{total}} = \dfrac{V}{R_1} + \dfrac{V}{R_2} + \dfrac{V}{R_3}$

So $\dfrac{1}{R_{total}} = \dfrac{1}{R_1} + \dfrac{1}{R_2} + \dfrac{1}{R_3}$

Therefore, for resistors in parallel the total resistance is the reciprocal of the sum of the reciprocals of the individual resistances.

Worked example

Find the total resistance of the following resistor networks.

(a)   560 Ω   1.5 kΩ   10 kΩ   **(1 mark)**

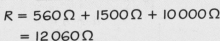

$R = 560\,\Omega + 1500\,\Omega + 10\,000\,\Omega$

$= 12\,060\,\Omega$

(b)   **(2 marks)**

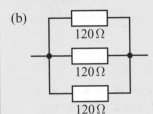

120 Ω

120 Ω

120 Ω

$\dfrac{1}{R} = \dfrac{1}{120\,\Omega} + \dfrac{1}{120\,\Omega} + \dfrac{1}{120\,\Omega} = \dfrac{3}{120\,\Omega}$

Therefore $R = 40\,\Omega$

(c)   **(2 marks)**

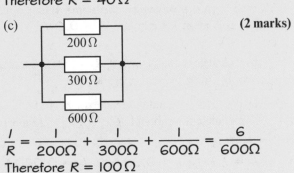

200 Ω

300 Ω

600 Ω

$\dfrac{1}{R} = \dfrac{1}{200\,\Omega} + \dfrac{1}{300\,\Omega} + \dfrac{1}{600\,\Omega} = \dfrac{6}{600\,\Omega}$

Therefore $R = 100\,\Omega$

> Remember to work in consistent units.

> If your answer is not smaller than the smallest value of the resistors in parallel then your answer is wrong!

Now try this

Find the total or equivalent resistance of the following resistor combinations.

(a)   **(3 marks)**

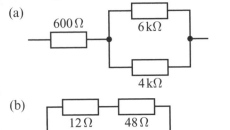

600 Ω   6 kΩ

4 kΩ

(b)   **(3 marks)**

12 Ω   48 Ω

90 Ω

Resistivity

Resistance is a property of a component; resistivity is a property of a material.

Resistivity

The resistance of a component such as a wire can easily be measured, and depends on the wire's:

- length, l
- cross-sectional area, A
- the **resistivity**, ρ, of the material from which it is made.

Resistivity is a property of materials and, at constant temperature, it does not vary with the size of a sample of material, whether you measure it for a tiny wire or a huge block. It is measured in $\Omega\,m$. Resistivities of metals are all of a similar order, around $10^{-8}\,\Omega\,m$.

Experiment shows that, for a resistor like a length of uniform wire,

$$R \propto l \text{ and } R \propto \frac{l}{A}$$

When these expressions are combined, the constant of proportionality is the resistivity ρ of the material at that temperature.

$$R = \frac{\rho l}{A}$$

Resistivity changes with temperature.

Relating resistance and resistivity

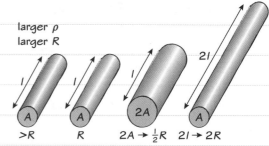

larger ρ
larger R

l l l $2l$
A   A   2A   A
>R   R $2A \to \frac{1}{2}R$ $2l \to 2R$

$$R = \frac{\rho l}{A}$$

Changes to resistance R if:

- length is changed:
  doubling the length to $2l$ gives
  $$\frac{\rho 2l}{A} = 2R: \text{ resistance also doubles}$$

- area is changed:
  doubling the width changes the area $A = 2\pi r^2$
  to $2\pi \times 2^2 \times r^2 = 4A$, which gives $\frac{\rho l}{4A} = \frac{R}{4}$:
  resistance drops to one-quarter of its previous value

- material is changed:
  using a conductor with twice the resistivity gives
  $$\frac{2\rho l}{A} = 2R: \text{ resistance also doubles.}$$

Worked example

Find the resistance of a 1.5 m length of wire of diameter 0.50 mm and resistivity $5.0 \times 10^{-7}\,\Omega\,m$. **(3 marks)**

area $A = \pi r^2 = \pi \times \left(\dfrac{0.50}{2} \times 10^{-3}\right)^2 m^2$

$R = \dfrac{\rho l}{A}$

$\quad = 5.0 \times 10^{-7} \times \dfrac{1.5}{\left(\pi \times \dfrac{0.50}{2} \times 10^{-3}\right)^2}$

$\quad = 3.8\,\Omega$ to 2 s.f.

Take care when calculating areas when you are given the diameter in millimetres.

Worked example

The resistance of a 1.30 m length of wire is $0.8\,\Omega$. The average diameter is 0.40 mm. Calculate the resistivity of the material the wire is made from. **(3 marks)**

$A = \pi r^2 = \pi \times \left(\dfrac{0.40}{2} \times 10^{-3}\right)^2 m^2$

$R = \dfrac{\rho l}{A}$ so $\rho = \dfrac{RA}{l}$

$\quad = \dfrac{0.8 \times \pi \times \left(\dfrac{0.40}{2} \times 10^{-3}\right)^2}{1.30}$

$\quad = 7.7 \times 10^{-8}\,\Omega\,m$

Potential and distance

Potential drops around a circuit: there is a potential difference across each component, even simple wires (though we often neglect this). The potential along a uniform current-carrying wire must vary with the distance:

$$R = \frac{V}{I} = \frac{\rho l}{A}$$

For any given uniform wire carrying a given current, I, ρ and A are constants, so the p.d. $V \propto l$.

Now try this

1 A steel wire has a resistance of $12\,\Omega$. A wire three times as long and twice the diameter will have a resistance of:

   A $9\,\Omega$ **B** $18\,\Omega$ **C** $27\,\Omega$ **D** $36\,\Omega$ **(3 marks)**

2 Find the resistance of a 3.5 m length of copper wire of diameter 0.40 mm. The resistivity of copper is $1.7 \times 10^{-8}\,\Omega\,m$. **(3 marks)**

Resistivity measurement

Practical skills The resistivity of a material is defined as the value of the resistance between opposite faces of a cubic metre of the material.

Determining the electrical resistivity of a material

Apparatus:
- a 2 m length of wire made of the material you are investigating
- micrometer to measure wire diameter, d
- metre rule to measure wire length, l
- power supply (1–3 V), ammeter and voltmeter, or else ohmmeter to measure R.

length of wire under test

Measure:
- the diameter d of the wire in several places and take the mean
- the length l of the wire
- the resistance R of the length of wire with the ohmmeter or pass a current through the wire and find R using current and p.d. measurements ($R = V/I$).
- repeat for different lengths, or for the same lengths of wires but with different cross-sectional areas.

Pass current through the wire for a short time only, so that the wire does not get hot. This will cause a change in the resistivity as well as being a potential burn hazard.

Calculation: varying lengths

If different lengths of wire are used:

Plot a graph of resistance against length of wire for at least six different lengths. Draw the best-fit line.

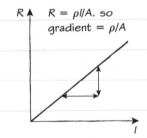

$R = \rho l/A$. so gradient $= \rho/A$

Measure the gradient of the line. Calculate A, the cross-sectional area of the wire using $A = \pi r^2$. Multiply the gradient by the value of A to find ρ.

Calculation: varying areas

If different cross-sectional areas of wire are used:

Plot a graph of resistance R against l/A for at least six different cross-sectional areas. Draw the best-fit line.

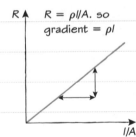

$R = \rho l/A$. so gradient $= \rho l$

Divide the gradient by the length of the wires used to find ρ.

Maths skills Straight-line graphs are characterised by the equation $y = mx + c$, where m is the gradient. The equation $R = \rho l/A$ follows this form, with $m = \rho/A$ (which is a constant here) and $c = 0$.

Now try this

In an investigation to find the resistivity of nichrome (an alloy of nickel, iron and chromium), the following resistances were recorded for nichrome wires of varying length and of cross-sectional area 0.2 mm² at 20°C.

(a) Plot a graph of the results and use it to calculate the resistivity of nichrome at 20°C.   **(6 marks)**

(b) Copper has a resistivity of $1.7 \times 10^{-8}\,\Omega\,\text{m}$ at 20°C. Comment on your result in the light of this and the fact that nichrome is used in heating elements.   **(1 mark)**

l /m	0.402	0.501	0.599	0.698	0.805	0.901	0.997	1.103
R /Ω	2.20	2.78	3.31	3.79	4.41	4.91	5.45	6.02

Current equation

Current is the rate of flow of charge, and we can derive an equation for it in the case of the movement of electrically charged particles through a conducting material.

The current equation

I is the current in amps through a conducting material of cross-section A.

This is called the **current equation** or transport equation:

$I = nqvA$

n is the **carrier density** of the material, the number of mobile charge carriers per cubic metre

q is the charge per carrier

v is the **mean drift velocity** of the charge carriers as they move through the conductor.

Worked example

Calculate the mean drift velocity of electrons in a copper cable with a cross-sectional area of $1.0\,mm^2$ passing a current of 5.0 A. (The carrier density for copper is 1×10^{29} electrons m^{-3} and the charge on an electron is $-1.60 \times 10^{-19}\,C$.) **(3 marks)**

$I = nqvA$

$v = \dfrac{I}{nqA}$

$\quad = 3.1 \times 10^{-4}\,m\,s^{-1}\ (0.31\,mm\,s^{-1})$

Less than a third of a millimetre per second!

First rearrange the equation to make v the subject.

Then remember to convert the area to m^2 from $mm^2 \rightarrow 1\,mm^2 = 10^{-6}\,m^2$.

You may ignore the sign of the charge here.

Carrier densities

The carrier density, n, in **metals** is very large. This is why metals are very good conductors and have very low values of resistivity.

Insulators have almost no mobile charge carriers and therefore their resistivities are huge.

Semiconductors, like silicon and germanium, have relatively low carrier densities. The value of n for silicon at room temperature is about a billion times smaller than copper. Therefore semiconductors have much higher resistivities than metals.

The effect of temperature on the resistivity of metals

The carrier density in metals is not affected by increase in temperature. However, the mean drift velocity, v, of the carriers is reduced by the increase in lattice vibrations. So the resistivity of metals increases with temperature.

At room temperature this lamp has a resistance of $10\,\Omega$; at its operating temperature, $\sim 2500\,°C$, the resistance is $100\,\Omega$.

Thermistors

Thermistors are made with semiconductor materials. In negative temperature coefficient (NTC) thermistors, as the temperature increases, the number of charge carriers also increases. The effect of increased lattice vibrations increasing resistivity is small in comparison, so the conductivity of the thermistor rises – its resistivity is reduced by the increase in n.

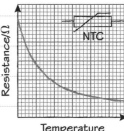

Light-dependent resistors (LDRs)

LDRs are also made of semiconductor materials. Energy falling on an LDR in the form of light frees more charge carriers, increasing the carrier density, so more light means lower resistivity.

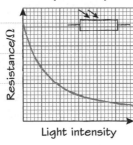

Now try this

1 A current of 2.0 A passes through a copper wire, 1. If a current of 4.0 A is passed through a second copper wire, 2, with twice the diameter, then the mean drift velocity of electrons in wire 2 is:
   A the same as in wire 1   **B** half that in wire 1
   C double that in wire 1   **D** 4 times that in wire 1
   (1 mark)

2 What assumption is necessary in order to answer question 1?   **(1 mark)**

E.m.f. and internal resistance

Cells transfer some energy internally when they are used, so not all the energy of the cell is transferred in the circuit.

Electrical cells

The **electromotive force (e.m.f.)**, $\mathscr{E}$, of a supply (such as a cell) is the energy gained per unit charge by charges passing through the supply. It is measured in volts ($1\,V$ is $1\,J\,C^{-1}$).

When a cell supplies a current to an external load, some of the energy is dissipated (transferred without doing useful work) within the cell itself because of the **internal resistance**, r, of the cell.

Symbol for a real electrical cell – the circles represent the physical terminals of the cell.

E.m.f., terminal p.d. and 'lost volts'

When a load, R, is connected to a cell, a current, I, is drawn from the cell. Some of the energy converted in the cell is dissipated within the cell itself.

terminal p.d.

$V = \mathscr{E} - Ir$

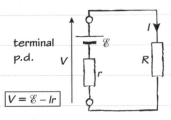

This results in the **terminal p.d.** V (the p.d. between the cell terminals) dropping by Ir, so $V = \mathscr{E} - Ir$. This p.d. drop across the cell's internal resistance is sometimes referred to as '**lost volts**'.

Finding the internal resistance of a cell

The current, I, delivered by the cell is varied by changing the value of the variable external resistor, R, and the terminal p.d., V, is noted for each value of I.

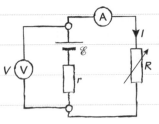

A graph of V against I will be a straight line that cuts the p.d. axis at $\mathscr{E}$ and has a gradient of $-r$.

Do not reduce R too much – you will short-circuit the cell!

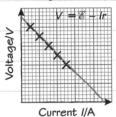

$V = \mathscr{E} - Ir$

Voltage/V

Current I/A

Worked example

The graph shows how the terminal p.d. of a battery of six cells varies with the current drawn from the battery.

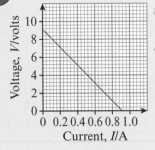

Voltage, V/volts

Current, I/A

(a) Determine the e.m.f. of one of the cells in the battery. **(2 marks)**

The intercept on the p.d. axis gives the e.m.f of the battery, $9\,V$. Therefore the e.m.f of one cell is $1.5\,V$, assuming the cells are identical and connected in series.

(b) Determine the internal resistance of one of the cells in the battery. **(2 marks)**

The gradient of the line is -10 so the internal resistance r of the battery is $10\,\Omega$, and of one cell is $1.7\,\Omega$.

Worked example

A cell has an e.m.f. of $1.5\,V$ and an internal resistance of $2.0\,\Omega$. Calculate the current drawn from this cell and its terminal p.d. when connected to load

(a) $R = 18.0\,\Omega$, (b) $R = 8.0\,\Omega$. **(4 marks)**

(a) Total resistance in the circuit $= (2 + 18)\,\Omega$

therefore $I = \dfrac{1.5}{20.0} = 0.075\,A$

Internal resistance p.d. drop $= 0.075 \times 2.0$
$= 0.15\,V$ so terminal p.d. $= (1.5 - 0.15)$
$= 1.35\,V$

(b) Similarly $I = \dfrac{1.5}{10.0} = 0.15\,A$

and terminal p.d. $= (1.5 - 0.3) = 1.2\,V$

Now try this

Two identical cells are connected together in parallel to form a battery, as shown.

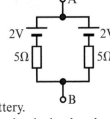

(a) A load of resistance $15\,\Omega$ is connected across the terminals A and B of this battery. Calculate the current, I, drawn by the load and the terminal p.d. V_{AB}. **(4 marks)**

(b) The cells are then connected in series. (i) What is the e.m.f. of this new battery and what is its internal resistance? (ii) What will the terminal p.d. of this battery be when the $15\,\Omega$ load is connected to it? **(4 marks)**

Potential divider circuits

Potential divider circuits share the voltage from a supply between two resistors and are used to provide a calculated fraction of the supply voltage.

The potential divider circuit

This circuit is used to provide a voltage V_{out} from a supply voltage V_{in}.

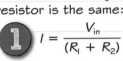

The current through each resistor is the same:

① $I = \dfrac{V_{in}}{(R_1 + R_2)}$

② $I = \dfrac{V_{out}}{R_1}$

Equating **①** and **②**:

$\dfrac{V_{out}}{R_1} = \dfrac{V_{in}}{(R_1 + R_2)}$ so $V_{out} = \dfrac{V_{in} \times R_1}{(R_1 + R_2)}$

This is the **potential divider equation**.

The output voltage V_{out} can be varied from 0 V to V_{in} by changing the resistance ratio.

$I = \dfrac{V_{out}}{R_1} = \dfrac{V_2}{R_2}$ therefore $\dfrac{V_{out}}{V_2} = \dfrac{R_1}{R_2}$

The voltage is split in the same ratio as the resistances.

In the LDR circuit right, output p.d. decreases with increasing light intensity. However, light sensors are often used to switch on lighting when needed. A circuit in which p.d. increases with light intensity would be more useful. This can be achieved by using the p.d. output across R_2.

A temperature sensor

The resistance of an NTC thermistor decreases as temperature increases, so it can be used in a potential divider as the basis of an electronic thermometer.

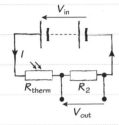

Thermistors are described on page 29.

$V_{out} = \dfrac{V_{in} \times R_2}{(R_{therm} + R_2)}$

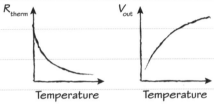

The output can be connected to a high-resistance voltmeter and calibrated to read temperature.

Detecting light intensity

As light intensity increases, the resistance of a light-dependent resistor (LDR) decreases. If the LDR is used in a potential divider, the output p.d. will vary with light intensity. LDRs are described on page 29.

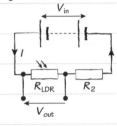

$V_{out} = \dfrac{V_{in} \times R_{LDR}}{(R_{LDR} + R_2)}$

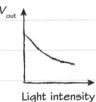

If a voltmeter is connected across the output it can be calibrated to read in lux (light intensity units).

Worked example

A thermistor is used in a potential divider circuit like the one above, with a 6.0 V supply. It is in series with a fixed resistor of resistance 1 kΩ. The resistance of the thermistor varies from 5.0 kΩ at 10 °C to 100 Ω at 80 °C.

Calculate the p.d. across the fixed resistor
(a) at 10 °C and (b) at 80 °C.   **(4 marks)**

(a) At 10 °C the resistance of the thermistor is 5000 Ω.

$V_{out} = \dfrac{V_{in} \times R_2}{(R_{therm} + R_2)}$

$V_{out} = \dfrac{6.0 \times 1000}{(5000 + 1000)} = 1.0\,V$

(b) At 80 °C the resistance of the thermistor is 100 Ω.

$V_{out} = \dfrac{V_{in} \times R_2}{(R_{therm} + R_2)}$

$V_{out} = \dfrac{6.0 \times 1000}{(100 + 1000)} = 5.5\,V$

Now try this

An LDR is used as a light sensor in such a potential divider circuit. Its resistance varies from 50 Ω in bright light to 2.5 kΩ in the dark. The supply p.d. is 5.0 V. The output p.d. is taken across the fixed resistor in series with the LDR.

Calculate the range of p.d. outputs when the fixed resistor is:
(a) 25 Ω   **(3 marks)**
(b) 500 Ω   **(3 marks)**
(c) 10 kΩ   **(3 marks)**

Exam skills 3

This exam-style question uses knowledge and skills you have already revised. Have a look at pages 23–31 for a reminder about how to analyse DC electric circuits.

Worked example

The circuit of fig. 1 consists of a direct current supply of e.m.f. 24 V, negligible internal resistance and three resistors.

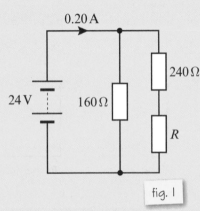

fig. 1

Two of the resistors have resistances 160 Ω and 240 Ω as shown.

The current drawn from the supply is 0.20 A.

(a) Calculate the resistance of R. **(3 marks)**

The current through the 160 Ω resistor is

$I = \dfrac{24}{160} = 0.15$ A.

The current through the other two resistors is therefore 0.20 − 0.15 = 0.05 A.

The total resistance of the two resistors on

the right is $R_{TOT} = \dfrac{24}{0.05} = 480\,\Omega$.

Therefore R = 480 − 240 = 240 Ω

(b) Resistor R is now short-circuited by connecting a wire of negligible resistance in parallel with it.

State and explain what happens to the currents in each arm of the circuit when R is short-circuited. **(3 marks)**

There is no change in the current through the 160 Ω resistor. This is because it still has a potential difference of 24 V across it.

The current is $\dfrac{24}{160} = 0.15$ A.

The current through the 240 Ω resistor increases. This is because the resistance of this arm has reduced but the potential difference across it has remained the same (24 V).

The current rises to $I = \dfrac{24}{240} = 0.10$ A.

Remember that, for conservation of charge, since 0.20 A enters this junction, the sum of currents leaving it must also be 0.20 A. This will be useful when you work out the currents in the parallel arms.

The two branches of this circuit are in parallel with each other, so they have the same potential difference across them (24 V). You can use this together with Ohm's law to calculate the current in each parallel arm.

Start by calculating the current through the 160 Ω resistor. This is connected directly across the supply, so it has a p.d. of 24 V across it.

This is where the charge conservation is used (see page 25, for a reminder of this important circuit law).

Notice how each step in the calculation has been presented in sequence and explained clearly – this helps to avoid making careless errors, and also shows that you fully understand what you are doing!

A short circuit has zero resistance. All current will flow through the short-circuit wire and not through R. The effect is the same as removing R and replacing it with the wire alone.

Command word: 'explain'

If a question asks you to **explain** something, you should:

- ✓ write in full sentences
- ✓ explain each step in the process
- ✓ use correct scientific language.

Density and flotation

Whether an object will float or sink in a fluid depends on the densities of the object and the fluid.

Density, ρ

The **density** of an object is defined by the equation:

$$\rho = \frac{m}{V}$$

where m is the mass of the object and V is its volume. S.I. units: $kg\,m^{-3}$.

Worked example

A block of iron measures $10\,cm \times 5.0\,cm \times 3.0\,cm$ and has a mass of $1.185\,kg$.
Find its density. **(2 marks)**

Volume in m³ = $(0.1 \times 0.05 \times 0.03)\,m^3$

$$\rho = \frac{m}{V} \text{ so } \rho = \frac{1.185\,kg}{(0.1 \times 0.05 \times 0.03)m^3}$$

$$= 7900\,kg\,m^{-3}$$

Worked example

(a) Describe how you would find the density of a liquid. **(3 marks)**

Determine the mass of a measuring cylinder empty then pour in the liquid, noting its volume, V. Find the total mass and from there the mass, m, of the liquid and substitute into the formula.

> Remember to read the volume marking at the level of the meniscus to avoid parallax error.

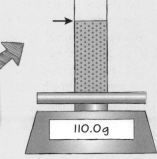

110.0g

(b) A measuring cylinder has a mass of 60 g empty and 110 g when filled with 50 cm³ of water. Calculate the density of water. **(3 marks)**

The mass of water is $\frac{110 - 60}{1000} = 0.050\,kg$

The volume of water is $5.0 \times 10^{-5}\,m^3$

$$\rho = \frac{m}{V} = \frac{0.050}{5.0 \times 10^{-5}} = 1000\,kg\,m^{-3}$$

The density of water is often given as $1\,g\,cm^{-3}$
$1000\,kg = 10^6\,g$ and $1\,m^3 = 10^6\,cm^3$

Upthrust and flotation (Archimedes' principle)

When an object is wholly or partly immersed in a liquid, the liquid exerts an upward force on the object called the **upthrust**. This can be shown by a simple experiment:

In air, a block is found to have a weight of $0.59\,N$. When it is wholly immersed in water it appears to have a weight of only $0.20\,N$.

This is because the water displaced by the block is exerting an upthrust on the block **equal to the weight of the displaced water**.

The free body diagrams show the forces acting on the block in each situation.

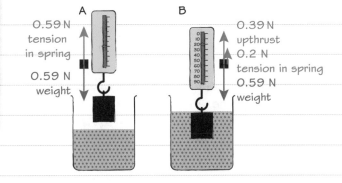

A
0.59 N tension in spring
0.59 N weight

B
0.39 N upthrust
0.2 N tension in spring
0.59 N weight

Now try this

A block of wood measuring $10\,cm \times 8\,cm \times 5\,cm$ has a mass of $240\,g$.

(a) Find the density of the wood in (i) $g\,cm^{-3}$ and (ii) $kg\,m^{-3}$. **(2 marks)**

(b) The block of wood is placed in a deep bowl of water, where it floats.
  (i) Explain why it floats. **(1 mark)**
  (ii) Calculate the percentage of the volume of the block above the water line.
    (Density of water = $1000\,kg\,m^{-3}$.) **(4 marks)**

Worked example

Will the block described above right float or sink in the water if released from the spring balance? **(2 marks)**

The upthrust has the maximum value for the block because the block is wholly immersed. As the weight of the block is greater than the upthrust, the block will sink when released from the spring balance.

Viscous drag

Viscous drag is the force that opposes the motion of an object through a fluid.

Stokes's law

The force, F, opposing the motion of a **spherical** object moving through a fluid is given by

$$F = 6\pi\eta rv$$

where η is the **viscosity** (or coefficient of viscosity) of the fluid (unit $N\,m^{-2}\,s$ or $Pa\,s$) r is the radius of the object and v is the velocity of the object through the fluid.

This equation is only valid if:
- the object is small, spherical and moving slowly
- if the flow of fluid around the object is **laminar**.

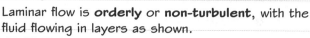

motion   DRAG

Laminar flow is **orderly** or **non-turbulent**, with the fluid flowing in layers as shown.

Effect of temperature on viscosity

The viscosity of most liquids decreases with temperature. For example, warming honey makes it runnier, i.e. less viscous. Car engine oil is more viscous at low temperature, which is one reason car engines may be more difficult to start in winter. Temperature of the liquid should be noted when determining its viscosity.

Worked example

Calculate the net downward force, W_N, on a steel sphere of volume $V = 1.1 \times 10^{-7}\,m^3$ when wholly immersed in water. (density of steel $\rho_s = 7900\,kg\,m^{-3}$, density of water $\rho_w = 1000\,kg\,m^{-3}$, $g = 9.81\,N\,kg^{-1}$.)   **(3 marks)**

Weight = mg
$= \rho Vg$ so the weight of steel sphere
$= \rho_s Vg$ and the upthrust
= weight of displaced water = $\rho_w Vg$

Since W_N = weight − upthrust

$W_N = (\rho_s - \rho_w)Vg$

so $W_N = (7900 - 1000) \times 1.1 \times 10^{-7} \times 9.81$
$= 7.4 \times 10^{-3}\,N$

Worked example

Glycerin has a coefficient of viscosity, η, of 1.42 Pa s at 20 °C. Calculate the drag force on a ball of diameter 1.0 cm falling at 20 cm s⁻¹ through glycerin.   **(2 marks)**

$v = 0.20\,m\,s^{-1}$, $r = 5.0 \times 10^{-3}\,m$

$F = 6\pi\eta rv$

$F = 6\pi \times 1.42 \times 5.0 \times 10^{-3} \times 0.20$
$= 0.027\,N$

🧪 Practical skills — Determining the viscosity of a liquid by a free-fall method

[0·000]

Fill a long cylinder with the liquid whose viscosity is to be determined.

Mark lines on the cylinder at regular distances. Measure the distance, x, between these lines.

Release a small sphere of known diameter, d, at the top of the column. Use a stopwatch to measure the time taken for the ball to fall through each interval, t_{AB}, t_{BC} and t_{CD}.

The times will be the same, and a mean can be taken, if the ball has reached terminal velocity.

If $t_{BC} > t_{AB}$ then the ball is still accelerating and the experiment must be repeated with a longer column of liquid or a smaller ball.

The ball reaches terminal velocity when: the net force downwards = drag (force upwards),

$W_N = 6\pi\eta rv$

$r = \dfrac{d}{2}$, $v = \dfrac{x}{t_{CD}}$ and the net downward force, W_N, on the ball when it is wholly immersed is found by calculation or by weighing the ball when wholly immersed in the liquid.

Hence the viscosity of the liquid may be found.

Now try this

1 A glass marble has a radius of 7.0 mm and is made of glass of density 2500 kg m⁻³. It is falling through a liquid, of density 1200 kg m⁻³ and coefficient of viscosity 0.50 Pa s, at a velocity of 10 cm s⁻¹.
   (a) Find the (i) volume, (ii) mass and (iii) weight of the marble.   **(3 marks)**
   (b) Find the upthrust on the marble.   **(2 marks)**
   (c) What is the drag force on the marble?   **(2 marks)**

2 A sphere is falling through a liquid at its terminal velocity. Describe what will happen to it if the temperature of the liquid suddenly drops.   **(2 marks)**

Hooke's law

The extension of an object is proportional to the force applied (provided that its elastic limit is not exceeded).

Hooke's law

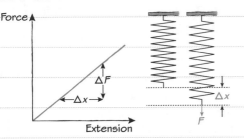

A force, F, applied to an extendable object, like a spring, causes an extension, Δx. **Hooke's law** states that the extension is proportional to the force applied, provided the object is not overstretched. This is **elastic deformation**. The size of the extension for a given force depends on the **stiffness**, k, of the object.

$$\Delta F = k\Delta x$$

Features of force–extension graphs

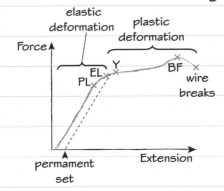

PL: the **limit of proportionality**. The object obeys Hooke's law up to this point.

EL: the **elastic limit**. The object will not return to its original length if stretched beyond this point; it will have a permanent set when the force is removed.

Y: the **yield point**. The extension increases fast for little increase in force as **plastic deformation** begins. In many materials EL and Y are the same.

BF: The **breaking force**. Once this point is reached the object will break even if the force is reduced.

A light spring is subjected to a tensile force of 6.0 N, which extends the spring by 120 mm.

(a) Calculate the stiffness of the spring. **(2 marks)**

$$k = \frac{\Delta F}{\Delta x} = \frac{6.0}{0.12} = 50\,\text{N m}^{-1}$$

(b) A force of 10 N is now applied to the spring. Find the new extension assuming Hooke's law is obeyed. **(2 marks)**

$$\Delta x = \frac{\Delta F}{k} = \frac{10}{50} = 0.20\,\text{m}$$

Force–compression graphs

Springs are frequently used in **compression**, for example in bathroom scales and car suspensions. A force–compression graph looks much the same as a force–extension graph.

> Generalised force–extension graph for a material in **tension** (stretched). Not all objects (wires, springs, etc.) will exhibit all the features shown.

Elastic strain energy

When an object is stretched by a tensile force or compressed by a compressive force work is done on the object. This is stored as **elastic strain energy**, ΔE_{el}.

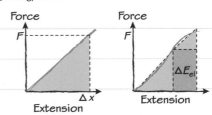

$\Delta E_{el} = \frac{1}{2}F\Delta x$, the area under the graph.

If the extension is non-linear, estimate ΔE_{el} by dividing the area into regular geometric shapes.

A spring with stiffness $k = 40\,\text{N m}^{-1}$ is extended by a load of 5.0 N. Calculate the extension which results and hence the elastic strain energy stored in the stretched spring. **(2 marks)**

$$\Delta x = \frac{\Delta F}{k} = \frac{5.0}{40} = 0.125\,\text{m}$$

$$\Delta E_{el} = \frac{1}{2}F\Delta x \text{ so } \Delta E_{el} = \frac{1}{2} \times 5 \times 0.125 = 0.31\,\text{J}$$

1 Explain the difference between the limit of proportionality and elastic limit of a wire in tension. **(2 marks)**

2 Two springs, one with stiffness $k = 100\,\text{N m}^{-1}$ and one with stiffness $k = 200\,\text{N m}^{-1}$, are joined end to end. A load of 20 N is suspended from this. Calculate the total extension and the elastic strain energy stored in each spring. **(4 marks)**

Young modulus

The Young modulus is a property of a material (rather than of an object made from that material) that is in tension or compression.

Stress

Stress is the force acting per unit cross-sectional area on a material.

Stress $\sigma = \dfrac{F}{A}$

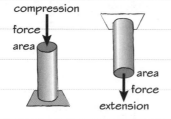

Stress is measured in $N\,m^{-2}$ or Pa, the same units as pressure.

Materials have a **breaking stress** that is the limit of the force per unit cross-sectional area that may be applied without the material failing.

Strain

Strain is the extension per unit length of material.

Strain

$\varepsilon = \dfrac{\Delta x}{x}$

Note that strain has no units as it is the ratio of two lengths.

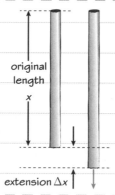

Young modulus, *E*, and stress–strain graphs

The **Young modulus**, E, is a measure of the stiffness of a material. It is the ratio of stress to strain for a material that is behaving elastically:

$E = \dfrac{\sigma \text{ STRESS}}{\varepsilon \text{ STRAIN}}$

This can be written

$E = \dfrac{F}{A} \times \dfrac{x}{\Delta x}$

The unit of the Young modulus is the pascal (Pa).

Determining the Young modulus

🧪 **Practical skills**

of a material

Two lengths of the wire to be tested are used: A is a reference and is kept in tension by a fixed load; B is the sample under test. The length l is measured with a tape. The diameter, d, is measured with a micrometer in several places and the mean value found. A range of loads are applied and the resulting extension for each measured with a Vernier scale.

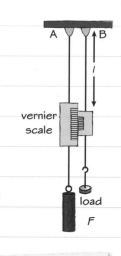

Typical apparatus for determination of the Young modulus.

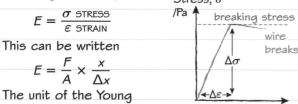

F /N	$\dfrac{\Delta x}{/10^{-4}\,m}$	$\sigma = F/A$ /MPa	$\varepsilon = \Delta x/x$

A graph of stress against strain is plotted, and the Young modulus found by measuring the gradient of the straight line graph: $\Delta\sigma/\Delta\varepsilon$

Now try this

A cable made of steel has a diameter of 18 mm and length 15 m. The Young modulus for steel is 200 GPa.

The cable is used to support a weight of 10 kN. Find
(a) the tensile stress in the cable under this load;
(b) the extension of the cable under load;
(c) the strain of the cable under load. **(3 marks)**

The UTS (ultimate tensile strength) of a material is another term for its breaking stress. The UTS of steel is 400 GPa.

(d) What is the maximum load that this cable can support before it fails? **(3 marks)**
(e) Find the minimum diameter of steel cable that could just support the 10 kN without breaking. **(3 marks)**

Worked example

The Young modulus for steel is 2.0×10^{11} Pa. Calculate the extension when a load of 50 N is applied to a 2.5 m length of wire with a cross-sectional area of $4.0 \times 10^{-7}\,m^2$. **(2 marks)**

$E = \dfrac{F}{A} \times \dfrac{x}{\Delta x}$

$\Delta x = \dfrac{Fx}{AE} = \dfrac{50}{4.0 \times 10^{-7}} \times \dfrac{2.5}{2.0 \times 10^{11}}$

$= 1.6 \times 10^{-3}\,m$ (1.6 mm)

Exam skills 4

This exam-style question uses knowledge and skills you have already revised. Have a look at page 36, for a reminder about stress, strain and the Young modulus.

Worked example

A student carries out an experiment to measure the Young modulus of iron. He sets up his apparatus as shown in the figure below, using an iron wire. He then measures the length of the wire for a range of different loads. He also measures the diameter of the wire.

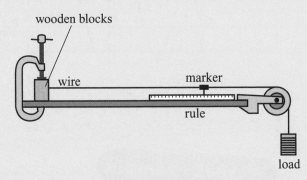

wooden blocks

wire

marker

rule

load

(a) Suggest a suitable piece of equipment for measuring the diameter *d* of the wire. Explain why this equipment is suitable. **(2 marks)**

A micrometer screw gauge is suitable because it can measure the diameter to the nearest 0.01 mm. This level of resolution reduces the uncertainty in this measurement.

(b) Explain why using a longer wire could reduce the uncertainty in the final value of the Young modulus. **(2 marks)**

A longer wire will extend more under the same load. The measurement error will not change but it will be a smaller fraction of the actual measurement so the percentage uncertainty in the result will be reduced.

(c) The student decides to plot a graph of stress against strain.

(i) Explain how he can calculate the stress σ in the wire for a load of mass *m*. **(2 marks)**

Stress = force/area, so he needs to divide the force provided by the load, $F = mg$, by the cross-sectional area of the wire, $A = \frac{1}{4}\pi d^2$: $\sigma = 4mg/\pi d^2$.

(ii) Explain how he can calculate the strain in the wire for any particular load. **(2 marks)**

Strain is defined as extension/original length. To find the extension he must subtract the original length from the extended length at this load, $e = l - l_0$.

(d) Explain how the student can calculate the Young modulus from the graph of stress against strain. **(2 marks)**

From the equation $\sigma = E\varepsilon$, a graph of stress σ (y-axis) against strain ε (x-axis) will be a straight line if the wire obeys Hooke's law. In this linear region the Young modulus E is the gradient of the graph, so the student must calculate the gradient.

Practical skills Make sure you know how to use a micrometer screw gauge to measure the diameter of a wire. Repeat the measurement at least three times, at different points along the wire and at different orientations, and use the average value. This will reduce the effect of random measurement errors.

Remember that the precision of the measuring instrument is equal to the smallest scale division on the instrument (in this case 0.01 mm).

Look at page 2 to review how measurement errors affect the uncertainty in your results.

Maths skills Be careful here – it is very easy to use the diameter instead of the cross-sectional area or to mix up diameter and radius!

Practical skills When working out extensions always subtract the original unstretched length from the loaded length. It is easy to think that the extension is just the extra length after adding one more mass to the hanger.

If you are doing this experimentally don't forget to include units. The gradient is stress divided by strain, so the units are $N\,m^{-2}$. Strain is dimensionless.

Waves

Waves transfer energy between places without the transfer of matter.

Terms

There are two types of wave: **longitudinal**, like sound waves, and **transverse**, like the ripples on the surface of water. Many of them are **mechanical waves** that require a medium through which to travel. **Electromagnetic waves**, like light, do not. All types have common features, which are described here in terms of mechanical waves:

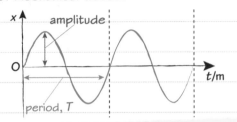

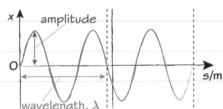

This graph shows how the displacement, x, of a particle in a medium changes with **time** t as a simple periodic wave passes through the medium.

This graph shows how the displacement x of particles along the line of travel of a simple periodic wave varies with **distance** s along the line of travel at an instant in time.

The **amplitude** of a wave is the maximum displacement of particles from their undisturbed position.

The **period T** of a wave is the time for one complete **cycle** or **oscillation** of the wave.

The **wavelength** λ of a wave is the distance between corresponding points on successive cycles of the wave.

The **frequency f** is the number of cycles of the wave passing through a point in the medium in one second. Frequency is measured in hertz, Hz ($1\,\text{Hz} = 1\,\text{s}^{-1}$).

The **speed v** of a wave is the distance travelled by a point on the wave in unit time. S.I. unit $\text{m}\,\text{s}^{-1}$.

The frequency f of a wave can be found from the period T.

$$f = \frac{1}{T}$$

The product frequency × wavelength is equal to the speed of waves through the medium.

$$v = f\lambda$$

Worked example

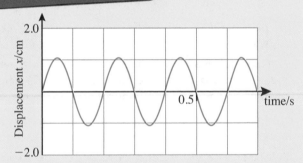

State the amplitude and period of the wave shown here, and calculate the frequency of the wave. **(3 marks)**

Amplitude: 1.05 cm, period: 0.2 s,

frequency $= \dfrac{1}{T} = 5.0\,\text{Hz}$.

Now try this

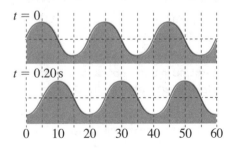

The two diagrams show the position of a ripple travelling across the surface of a tank of water at two instants in time. The lower one shows the water profile 0.20 s after the upper one.

Find (a) the wavelength, (b) the speed and (c) the frequency of the ripple. **(3 marks)**

Worked example

(a) Ripples travel across the surface of a pond at $15\,\text{cm}\,\text{s}^{-1}$. If the frequency of the ripples is 6.0 Hz find their wavelength. **(2 marks)**

$v = f\lambda$ so $\lambda = \dfrac{v}{f} = \dfrac{0.15}{6} = 0.025\,\text{m}$

(b) State what happens to the wavelength if the frequency is doubled. **(1 mark)**

From the formula it can be seen that doubling the frequency will halve the wavelength to 0.0125 m.

Longitudinal and transverse waves

Longitudinal waves propagate by the oscillation of particles displaced parallel to the line of travel of the wave. In transverse waves the particles are displaced at right angles to the line of travel.

Longitudinal waves

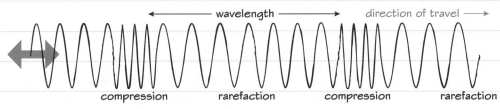

Longitudinal waves may be demonstrated on a 'Slinky' spring. Moving the end of the stretched spring backwards and forwards causes waves to travel along the spring as a series of **compressions** and **rarefactions**. The coils of the spring are displaced **parallel** to the direction of travel of the pulse of energy along the spring.

Sound waves travel through materials in the same way. A vibrating object causes **pressure variations** by pushing molecules closer together as it oscillates.

Worked example

Given a graph of displacement for particles against their distance from the start of the medium, plot a graph to show how the pressure in a longitudinal wave varies with distance along the medium. **(4 marks)**

The arrows show the size of the displacement of particles at various distances along the medium. Below this the original particle positions are shown as open circles and their new displaced positions are shown as black circles.

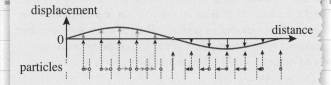

From this it can be seen that a compression occurs in the middle, where the central particle is not displaced, meaning that the pressure here is above the average, and the ends of the medium are regions of rarefaction and thus lowered pressure. The required graph is therefore:

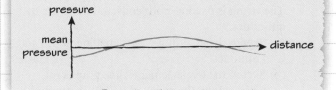

Transverse waves

Transverse waves are familiar from water surfaces. They can be created on a stretched spring or rope by moving the end of the rope at right angles to the rope.

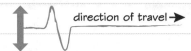

So in a transverse wave the particles of the medium carrying the wave energy are displaced perpendicular to the direction that the wave is travelling.

Now try this

Classify the following as transverse waves, longitudinal waves, or not waves.

(a) a bat's sonar chirps **(1 mark)**

(b) the rotation of a car wheel on the road **(1 mark)**

(c) the ripples on a cup of tea when the table is jogged **(1 mark)**

(d) the movement of air in an organ pipe producing a note **(1 mark)**

(e) the vibration of a guitar string when it is plucked **(1 mark)**

(f) the movement of a drum skin when it is hit **(1 mark)**

(g) the movement of a tree shaken by the wind **(1 mark)**

(h) the noise of a dentist's rotating drill. **(1 mark)**

Standing waves

Standing wave patterns can result from the superposition of two oppositely directed travelling waves.

Standing waves in a pipe

Wind instruments produce longitudinal **standing waves** or **stationary waves** in the air within them. Blowing across the end of a closed pipe produces a pressure variation that makes a longitudinal wave travel along the pipe. This wave is reflected back along the pipe and a standing wave is formed as the original and reflected waves **superpose**, i.e. add together. At a closed end of a pipe, the air particles cannot vibrate, so there is always a **node** in the standing wave.

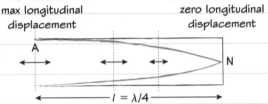

max longitudinal displacement   zero longitudinal displacement

$l = \lambda/4$

The **fundamental** or simplest standing wave with a node at the closed end and an **antinode** at the open end.

A standing wave in a pipe open at both ends has an antinode at each end.

Worked example

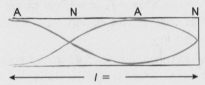

(a) Sketch the next simplest standing wave that can be set up in a pipe closed at one end. What is the wavelength of the sound in terms of the length, l, of the pipe? **(3 marks)**

There must be an antinode at the open end and a node at the closed end, so: $\lambda = \dfrac{4l}{3}$ (that is, $\frac{1}{3}$ of the fundamental wavelength).

A   N   A   N

$l =$

(b) State what happens to the frequency of the sound wave produced. **(1 mark)**

The speed of sound ($v = f\lambda$) is unchanged, so the frequency increases by a factor of 3.

Practical skills — Determining the speed of sound in air

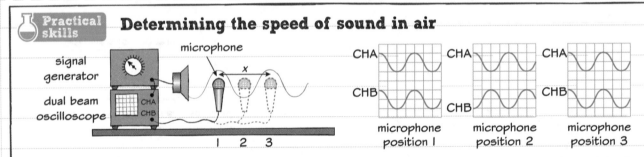

microphone

signal generator

dual beam oscilloscope

CHA
CHB

1 2 3

x

CHA   CHA   CHA
CHB   CHB
                CHB

microphone position 1   microphone position 2   microphone position 3

A signal generator is set to a known frequency f and connected to a loudspeaker and channel A of a dual-beam oscilloscope. A microphone is connected to channel B and is moved to a position, 1, in which the wave traces are in **phase**. The microphone is then moved onwards. The traces will become out of phase. (In position 2 the traces are in **antiphase**.) When the microphone has been moved through one wavelength λ the traces are back in phase (position 3). Since both f and λ are now known, the speed of sound in air may be calculated using $v = f\lambda$.

Standing waves on a stretched string

Travelling waves sent along a stretched string are reflected back when they reach the end, so two travelling waves are continuously moving along and superposing. If the string length is a whole number of half wavelengths, a standing wave results. At some points the two travelling waves will always be out of phase and will cancel out to form a node. Midway between adjacent nodes the two waves will always be in phase and will add to cause maximum movement of the string, or antinodes.

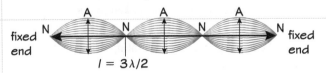

fixed end   N   A   N   A   N   A   N   fixed end

$l = 3\lambda/2$

Now try this

1 In an experiment to find the speed of sound in air, a signal frequency of 400 Hz was used and the distance between adjacent in-phase positions of the microphone was found to be 85 cm. Calculate the speed of sound in air. **(3 marks)**

2 A standing wave is set up on a string 3 m long. The string is fixed at both ends and there are two antinodes.
   (a) Sketch the standing wave, labelling nodes and antinodes. **(2 marks)**
   (b) Sketch the displacement–time graphs for points along the string at 0.75 m, 1.0 m and 2.0 m from one fixed end. **(3 marks)**

Phase and phase difference

When two waves are superposed, they interfere in a way that depends on their phase difference.

Phase angles and phase difference

A particle travelling around a circle at a steady speed and a particle oscillating up and down in a transverse wave look the same if seen side-on. This gives us a way to describe the position of a particle at a point in the wave cycle: the **phase angle**. The **phase** between a particle at position A (phase angle = 90°) and one at position B (phase angle = 270°) is thus 180°, or π radians.

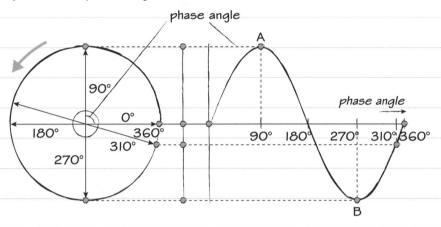

Worked example

This graph shows how the displacement x of particles in a medium varies with distance s from a travelling wave source at an instant in time.

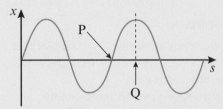

Sketch graphs to show how the displacement of a particle at (a) P and (b) Q varies with time. **(4 marks)**

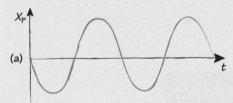

(a)

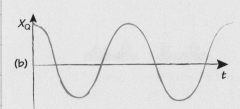

(b)

(c) What is the phase relationship between the wave at P and at Q? **(2 marks)**

Wave at Q leads P by 90° or $\dfrac{T}{4}$ (where T is the period).

Wavefronts

A **wavefront** is a line joining all points in a medium where the waves are in phase. Ripples on the surface of water in a shallow tank are often used to demonstrate the idea of wavefronts.

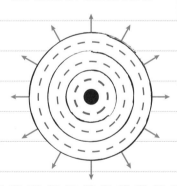

Circular wavefronts are set up by regularly disturbing the water in the centre. The blue lines show the crests of waves. The green dotted lines midway between the crests also join points in phase – the troughs. The troughs are exactly half a wavelength out of phase with the crests.

Now try this

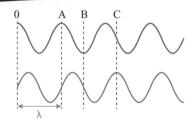

(a) State the phase difference between 0 and A, A and B, and B and C in degrees and in radians. **(3 marks)**

(b) State the phase difference between the top (green) wave and the lower (blue) wave in degrees and in terms of the period T. **(2 marks)**

Superposition and interference

Two or more waves occupying the same position in space are said to superpose.

Waves pass through one another

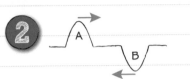

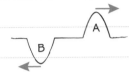

We can demonstrate this on a stretched rope. Here we show two cases of two wave pulses travelling towards each other along the rope and passing through each other unaffected.

Waves superpose

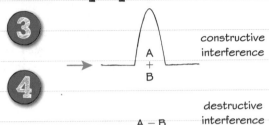

constructive interference

destructive interference

At the moment when the two waves occupy the same position along the rope, the wave pulses **superpose** and **interfere** with each other, adding or subtracting as shown.

Interference in two dimensions

The same rules that apply for waves in one dimension, on a rope, also apply in two dimensions. A ripple tank allows us to see the effect of superposition of waves travelling in two dimensions.

The red dotted lines show lines of **constructive interference**. Along these lines the two sets of travelling circular waves superpose in phase because the **path difference**, the difference between the distances the two waves have travelled, from the two sources, S_1 and S_2, to any point on these lines is either zero or a whole number of wavelengths. Between the red lines are blue dotted lines showing where the sets of waves are always in antiphase, because the path differences are an odd number of half wavelengths, resulting in **destructive interference** and hence very little disturbance of the water surface.

Worked example

Two similar sets of circular ripples are made at S_1 and S_2 with a frequency of $5.0\,\text{Hz}$ and travel at $16\,\text{cm s}^{-1}$ across the surface of a ripple tank. X is a point on the water surface 8 cm away from the source S_1.

(a) What is the path length S_1X in terms of the wavelength, λ, of the ripples? **(3 marks)**

$\lambda = \dfrac{v}{f} = \dfrac{0.16}{5.0} = 0.032\,\text{m}$

$S_1X = \dfrac{0.080}{0.032} = 2.5\,\lambda$

(b) Suggest a value of S_2X such that the two sets of waves arrive at X in phase (i) in terms of λ, (ii) in cm. **(2 marks)**

For the two sets of ripples to arrive at X in phase the path difference ($S_2X - S_1X$) must be a whole number of λ (or zero, but this is clearly not the case), so (i) $S_2X = 3.5\,\lambda$, or (ii) $3.5 \times 0.032\,\text{m} = 0.112\,\text{m}$.

Coherence

The two waves described in the worked example are **coherent**. This means that they have:
- the same frequency
- an unchanging phase relationship
- similar amplitudes
- the same type (e.g. light waves).

White light produced from heated filaments is incoherent: bursts of waves (photons) are produced from individual atoms with no fixed phase relationship or constant frequency. Laser light is highly coherent.

Now try this

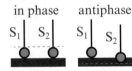

in phase   antiphase

In the worked example it was assumed that the dippers at S_1 and S_2 setting up the two sets of circular waves were in phase, moving up and down at exactly the same time.

Suppose the dippers move in antiphase, one starting to move down as the other starts to move up. Suggest and explain how this would affect what happens when the two sets of waves superpose at X. **(2 marks)**

Velocity of transverse waves on strings

For strings of the same length, the frequency of the fundamental standing wave depends on v.

The equation for v

The velocity, v (m s^{-1}), of transverse waves travelling along a string in tension is given by

$$v = \sqrt{\frac{T}{\mu}}$$

where T is the tension in the string in N and μ is the mass per unit length of the string in kg m^{-1}.

Worked example

On a standard guitar, the lowest-pitch string has a frequency of $f_1 = 82.4$ Hz and the highest has a frequency of $f_2 = 329.6$ Hz. The strings have the same tension and the same length. Find the ratio of their masses m_1 and m_2.   **(2 marks)**

The length, l, of the strings is $\frac{\lambda}{2}$ (distance between adjacent nodes).

Combining $f = \frac{v}{\lambda}$ and $\lambda = 2l$ with the velocity

equation gives $f = \frac{1}{2l}\sqrt{\frac{T}{\mu}}$

so $\dfrac{f_2}{f_1} = \dfrac{\sqrt{\mu_1}}{\sqrt{\mu_2}} = \dfrac{\sqrt{m_1}}{\sqrt{m_2}}$ as T and l

are the same for both strings.

$\dfrac{f_2}{f_1} = 4, \dfrac{m_1}{m_2} = 16.$

Processing the results of the investigation

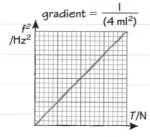

gradient $= \dfrac{1}{(4\,ml^2)}$

f^2 /Hz^2 T/N

To show how the fundamental frequency depends on the tension in the string, find the lowest (fundamental) frequency of resonance for the string over a range of different values of tension, keeping l and μ constant. Plotting a graph of f^2 against T should produce a straight line, as shown here, showing $f \propto \sqrt{T}$.

Repeating the experiment for strings of different lengths while keeping T and μ constant will show that $f \propto 1/l$, and repeating the experiment for strings of different mass/unit length while keeping T and l constant will show that $f \propto \dfrac{1}{\sqrt{\mu}}$.

Practical skills   **Investigating how l, T and μ affect f for a vibrating wire**

The simplest mode of vibration for a stretched string, which produces the lowest frequency vibration, has nodes at the fixed ends and a single antinode in the middle. Hence $\lambda = 2l$,

and $f = \dfrac{1}{2l}\sqrt{\dfrac{T}{\mu}}$ as shown in the worked example.

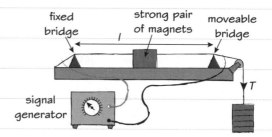

fixed bridge   strong pair of magnets   moveable bridge

signal generator

A sonometer, shown above, is used to investigate how f is affected by l, T and μ.

A measured length of the wire under test is weighed and μ, the mass per metre length, calculated. The tension, T, in the wire can be varied using known masses, and the length, l, of the wire can be varied by moving the bridge. An alternating current (AC) of known frequency is passed through the wire. Strong permanent magnet poles placed either side of the wire at the midpoint then make the wire vibrate. As the frequency of the AC is increased from zero, the wire will vibrate strongly when the resonant frequency, f, is reached, with an antinode at the centre of the wire and nodes at the fixed ends.

Just one of the independent variables, l, T and μ, is varied at a time and the resonant frequency found for different values.

Now try this

1. The bottom E string of a guitar is 64 cm long and at a tension of 75 N. The string vibrates at 82.4 Hz when plucked. What is the mass per unit length of the string?   **(3 marks)**

2. The pitch of the E string can be raised to that of the next string, the A string with $f = 110$ Hz, by shortening the string. What length should the E string be to vibrate at 110 Hz?   **(3 marks)**

The behaviour of waves at an interface

Waves can be transmitted and reflected on meeting an interface between media. The proportion of reflection to transmission depends on the medium on each side of the interface.

Pulse–echo techniques: sonar

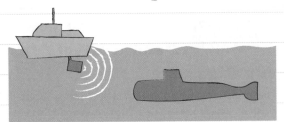

Undersea objects are located using sonar. Short pulses of sound are emitted and the time taken for each pulse to travel out and be reflected back Is measured, allowing the position of objects in the path of the sound to be determined.

Pulse–echo techniques: ultrasound scans

A gel is used to transmit sound from the **transducer** (the emitter, which transfers electrical energy into sound) through the outer surface of the body without reflection. Once inside the body, waves are reflected back from each different layer of tissue, making it possible to build a picture of, for example, an unborn baby.

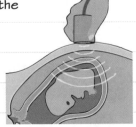

Pulse duration and wavelength

The worked examples show two applications of pulse–echo systems. The wavelengths

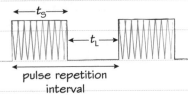

pulse repetition interval

used are quite different because the sizes of the objects being detected are very different. To detect small objects the wavelength must be much smaller because the **resolution** of the image formed is dependent on wavelength. To distinguish between two close points in an image a higher resolution is required and, therefore, the use of shorter wavelengths.

The **pulse duration** affects the **range** (maximum distance) over which the waves travel and return. The wave pulse is sent, and the echo needs to be detected before the next pulse is transmitted – within the **listening time**, t_L. Making this longer to increase the range reduces the amount of information that can be gathered.

Worked example

A sonar pulse–echo detection system uses sound with a frequency of 2.5 kHz. Sound travels at 1500 m s^{-1} in sea water.

(a) Find the wavelength of sound waves in sea water. **(2 marks)**

$$\lambda = \frac{v}{f} = \frac{1500}{2500} = 0.60\,\text{m}$$

(b) An echo is detected 2.4 s after the pulse is transmitted. Find the distance to the detected object. **(3 marks)**

Distance travelled = $v \times t$ = 1500 × 2.4 = 3600 m. This is the distance to the object and back, so the distance to the object is 1800 m.

Now try this

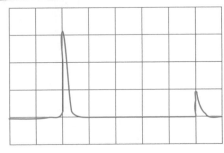

This oscilloscope trace shows the transmitted and received pulses used in a radar (radio detecting and ranging) range-finding system. Each square on the horizontal axis represents 20 µs.

(a) Find the distance to the detected object.
($c = 3.00 \times 10^8$ m s^{-1}) **(4 marks)**

(b) Explain why the reflected signal is smaller in amplitude than the transmitted signal. **(2 marks)**

Worked example

An ultrasound scanner produces sound waves with a wavelength of 0.50 mm. The waves travel through soft body tissue at 1540 m s^{-1}. An echo is detected from the fetus at a distance of 8.0 cm. (a) Find the frequency of the ultrasound transmitted. **(2 marks)**

$$f = \frac{v}{\lambda} = \frac{1540}{(5.0 \times 10^{-4})} = 3.08 \times 10^6\,\text{Hz} \ (3.1\,\text{MHz})$$

(b) Find the time interval between pulse and echo. **(3 marks)**

$$\text{Time} = \frac{2d}{v} = \frac{0.16}{1540} = 1.04 \times 10^{-4}\,\text{s} \ (100\,\text{µs})$$

Refraction of light and intensity of radiation

Refraction is the change in direction of a ray of light passing from one material to another at an angle.

Refraction of light in a glass block

The **refractive index (RI)** n_x for a medium x is defined as $n_x = \dfrac{c}{v_x}$, where c is the speed of light in a vacuum and v_x is the speed of light in medium x.

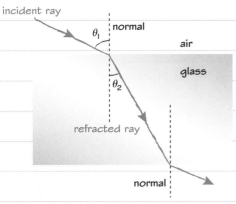

The path of a ray of light passing from medium 1 into medium 2 can be calculated using

$$n_1 \sin \theta_1 = n_2 \sin \theta_2$$

Since the speed of light in air is very close to that in a vacuum we can take n_1, the RI for air, to be 1.

Worked example

Light travels at $\frac{2}{3}c$ in ordinary glass.

(a) What is the RI of glass? **(1 mark)**

$$n_2 = c \div \frac{2c}{3} = 1.5$$

(b) If a ray of light travelling in water meets a boundary with glass at an angle to the normal of 50° what angle to the normal does the refracted ray make as it enters the glass? RI of water $n_1 = 1.33$. **(3 marks)**

$$n_1 \sin \theta_1 = n_2 \sin \theta_2$$
$$1.33 \sin 50° = 1.5 \sin \theta_2$$
$$\sin \theta_2 = \frac{1.33}{1.5} \sin 50°$$
$$\theta_2 = 42.8°$$

Note that the angles θ_1 and θ_2 are measured from the normal, as shown in the diagram. θ_1 is often labelled i (angle of incidence) and θ_2 is often labelled r (angle of refraction).

Practical skills — Measuring the RI of a solid

To measure the refractive index of a material, shine a narrow beam of light through a rectangular block of it. Draw around the block and mark the points B and C where the ray of light enters and leaves the block, and a point A on the incident ray. Remove the block and join A to B and B to C with straight pencil lines. Mark in the normal at B and measure the angle of incidence, θ_1, and the angle of refraction, θ_2.

Repeat for a range of values of θ_1 from 30° to 70°. Plot a graph of $\sin \theta_1$ against $\sin \theta_2$. Since $n_1 \sin \theta_1 = n_2 \sin \theta_2$, and n_1, the RI of air, can be taken to be 1, the gradient = n_2.

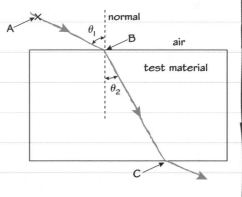

Intensity of radiation

The intensity of light, or any other form or radiated energy, is:

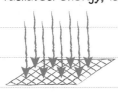

the energy arriving per second in watts ($1\,W = 1\,J\,s^{-1}$)

falling on an area of $1\,m^2$

$$\text{intensity (in W m}^2) = \frac{\text{power (in W)}}{\text{area (in m}^2)}$$

$$I = \frac{P}{A}$$

Now try this

1 Given that the speed of light in a vacuum is $3.00 \times 10^8\,m\,s^{-1}$ and the refractive index of water, n_w, is 1.33, calculate the speed of light in water. **(2 marks)**

2 The intensity of the Sun's radiation at the distance, R, of the Earth's orbit is about $1.4\,kW\,m^{-2}$. The radius of Saturn's orbit is about $10\,R$. Calculate the intensity of radiation reaching Saturn. **(3 marks)**

Total internal reflection

Under certain conditions, when light strikes a boundary between two media, all of the light is internally reflected.

Total internal reflection

Consider a light ray travelling from a semicircular glass block into air, as shown in the diagrams below. Because light travels faster in air than in glass, the ray bends away from the normal as it enters the air. Angle θ_2 increases as θ_1 is increased. When θ_1 reaches the **critical angle** the ray travels along the boundary at 90° (diagram 3). If θ_1 is increased further, **total internal reflection (TIR)** occurs (diagram 4).

Total internal reflection needs:
* light travelling in one medium at a boundary to another with a lower refractive index, i.e. from 'slow' to 'fast'
* light incident on the boundary at an angle greater than the critical angle.

$n_1 \sin \theta_1 = n_2 \sin \theta_2$, where n_1 is the refractive index (RI) of the material of the block and n_2 is the RI of the air, taken as 1.

Substituting the critical angle C for θ_1 and 90° for θ_2 gives $n_1 \sin C = 1$ or $\sin C = \dfrac{1}{n_1}$

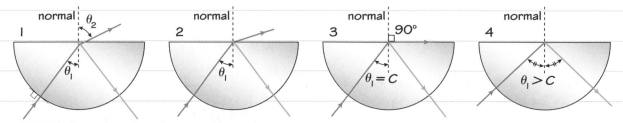

Usually when a ray of light meets a boundary between two different media, some light will pass across the boundary, undergoing refraction, and some will be reflected.

Worked example

The RI for ordinary glass is 1.5. The RI of diamond is 2.4. Calculate the critical angles for glass and diamond. **(2 marks)**

$\sin C = \dfrac{1}{n_1}$

For glass $n = 1.5$

$\therefore C = \sin^{-1}\left(\dfrac{1}{1.5}\right) = 41.8°$

For diamond $n = 2.4$

$\therefore C = \sin^{-1}\left(\dfrac{1}{2.4}\right) = 24.6°$

This smaller critical angle, together with the way diamonds are cut, accounts for how well they sparkle.

Worked example

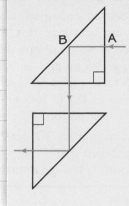

Total internal reflection is used in periscopes. The incident ray enters the system at A. Find the minimum value of the refractive index that the prisms must have. **(2 marks)**

The ray makes an angle of 45° at B.
Therefore $C \leq 45°$.

$n \geq \dfrac{1}{\sin C} \rightarrow n$ must be at least 1.4

Now try this

1 Optical fibres use total internal reflection to transmit information in digital form using pulses of light. One type of optical fibre consists of a core of refractive index n_1 and a cladding of refractive index n_2.

cladding
core
cladding

Select which of the following must be true for total internal reflection as shown to be possible:

   A $n_1 < n_2$ B $n_1 > n_2$ C $n_1 = n_2$ D $n_2 < 1$ **(1 mark)**

2 The tube on a stethoscope uses total internal reflection to transmit sound from the receiver to the doctor's ears. What does this suggest about the speed of sound in air compared to with the speed of sound in the plastic tube? **(1 mark)**

Exam skills 5

This exam-style question uses knowledge and skills you have already revised. Have a look at pages 38–46, for a reminder about waves and their properties.

Worked example

The figure shows a ray of light travelling from air into water. The refractive index of water is 1.33.

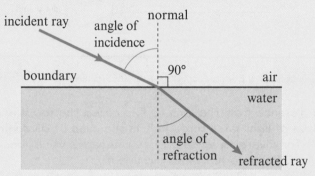

(a) Explain why light refracts at the boundary between air and water. **(2 marks)**

The speed of light in water is lower than the speed of light in air. Waves refract when they cross a boundary and the wave speed changes.

(b) The speed of light in air is $3.00 \times 10^8\ \mathrm{m\,s^{-1}}$.
Calculate the speed of light in water. **(2 marks)**

Speed of light in water $c_{water} = \dfrac{c}{n}$

$= \dfrac{3.0 \times 10^8}{1.33} = 2.26 \times 10^8\ \mathrm{m\,s^{-1}}$

> Quote the equation you are using before you substitute values.

(c) A particular ray has an angle of incidence equal to 70°.
Calculate the angle of refraction. **(3 marks)**

At the boundary:

$\dfrac{\sin i}{\sin r} = n$

$\sin r = \dfrac{\sin i}{n} = \dfrac{\sin 70°}{1.33} = 0.707$

Therefore $r = 45.0°$

> Make sure you start with a relevant equation and then show each stage of the working.

> Remember $n_1 \sin \theta_1 = n_2 \sin \theta_2$ (page 45). Here $n_1 = 1$, $n_2 = n$ the RI of water, $\theta_1 = i$ and $\theta_2 = r$

(d) Now consider rays of light travelling from water into air. Explain why there is a maximum angle of incidence, the critical angle, at the water/air boundary above which there will be no refracted ray. **(3 marks)**

Rays striking the water/air boundary refract away from the normal. This means that there will be an incident angle less than 90° for which the refracted angle will be 90°. For incident angles larger than this critical value there can be no refracted ray; instead, total internal reflection occurs.

> Here you might be tempted to simply say that total internal reflection will occur, but this doesn't answer the question. You are being asked **why** there is no refracted ray above a certain critical angle. Make sure your answer is focused on the question that is actually asked.

(e) Calculate the critical angle C at the water/air boundary. **(2 marks)**

$\sin C = \dfrac{1}{1.33} = 0.752$

critical angle $C = 49°$ (2 s.f.)

> Don't confuse C, for critical angle, with c, for speed of light!

Lenses and ray diagrams

Ray diagrams can be used to predict the images that lenses will create.

Types of simple lenses

converging lens   diverging lens

Parallel rays of light that meet a lens at right angles to the plane of the lens are converged or diverged as shown here.

F is called the principal **focus** of the lens, and the distance from the lens to F is called the **focal length**, f, of the lens. In the case of **converging** lenses, rays of light pass through F. In the case of **diverging** lenses rays of light only appear to originate from F. For diverging lenses, f has a negative value, because it is on the same side of the lens as the incident light. Rays do not pass through it.

Constructing ray diagrams

The location and nature of images formed by lenses can be found by drawing ray diagrams to scale with three rays whose paths are defined.

Converging lens

1 Rays that meet the plane of the lens at 90° pass through F.

2 Rays pass straight through the middle of the lens.

3 Rays that pass through F will emerge at 90° to the plane of the lens.

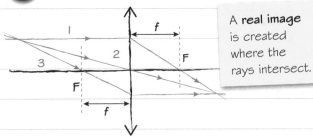

A **real image** is created where the rays intersect.

Diverging lens

1 Rays that meet the plane of the lens at 90° appear to have come from F.

2 Rays pass straight through the middle of the lens.

3 Rays travelling toward F on the other side of the lens emerge at 90° to the plane the lens.

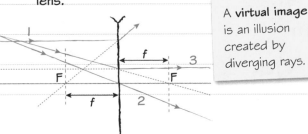

A **virtual image** is an illusion created by diverging rays.

Worked example

An object 10 cm tall is placed 30 cm from a converging lens of focal length 10 cm. Construct a ray diagram to determine the location of the image formed by the lens. Suggest a practical application.   **(4 marks)**

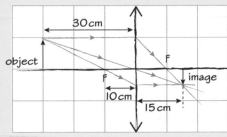

Draw ray diagrams to scale on graph paper. It is sufficient to draw just two rays to locate the image but all three will intersect at the image location. The diagram shows that the image is formed 15 cm from the lens, height 5 cm so diminished, and inverted (upside down). It is also a **real** image (the rays really intersect, they do not just appear to do so) so can be projected on a screen.

Application: lens in a simple camera.

Now try this

1 An object 8 cm high is placed 12 cm away from a converging lens of focal length 10 cm. Use a ray diagram to locate the image and suggest where a converging lens may be used in this way. **(5 marks)**

2 An object 5 cm tall is placed 10 cm from a converging lens of focal length 15 cm. Locate the image using a ray diagram and suggest a practical use.   **(5 marks)**

Lens formulae

The power and magnification of lenses and the magnification of their images can be calculated.

Power of a lens

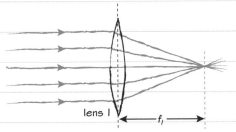

The power of a lens depends on its focal length. Lens 2 is more powerful than lens I.

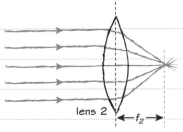

The power, P, of a lens is measured in dioptres (D) and is given by

$$P = \frac{1}{f}$$

where f is measured in metres.

Since f for a diverging lens is negative, its power is given as a negative number too.

Power of a compound lens

Many optical instruments use lenses combined into a compound lens. The power of such a lens is:

$$P = P_1 + P_2 + P_3 + \dots$$

The lens formula and 'Real is positive'

The location of an image can be found using the formula:

$$\frac{1}{f} = \frac{1}{u} + \frac{1}{v}$$

where f is the focal length of the lens, u is the distance between the object and the lens and v is the distance between the image and the lens.

Distances to real images are taken as positive.

Remember that f is negative for a diverging lens.

Worked example

A diverging lens has a focal length of 20 cm. An object 10 cm tall is placed 5 cm from the lens. Locate the image.   **(3 marks)**

This is a diverging lens, so $f = -0.20$ m.

The object distance $u = 0.05$ m.

$$\frac{1}{f} = \frac{1}{u} + \frac{1}{v} \rightarrow \frac{-1}{0.20} = \frac{1}{0.05} + \frac{1}{v}$$

$$\frac{1}{v} = \frac{-1}{0.20} - \frac{1}{0.05} \rightarrow \frac{1}{v} = \frac{-5}{0.20}$$

so $v = -0.04$ cm.

A virtual image is formed 4.0 cm from the lens on the same side as the object.

Magnification

To find the magnification, m, of an image:

$$m = \frac{h_1}{h_o}$$

where h_1 is the height of the image and h_o the height of the object. The ray diagrams below show that the ratio $\frac{h_1}{h_o} = \frac{v}{u}$ so magnification may be calculated from

$$m = \frac{v}{u}$$

The two shaded triangles are similar

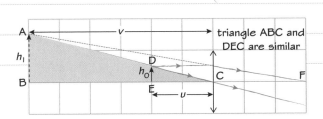

triangle ABC and DEC are similar

Worked example

A projector has a lens of focal length of 50 cm. A slide is placed 52 cm from the lens. Use the lens formula to find the image position and use this to calculate the magnification.   **(X marks)**

$$\frac{1}{f} = \frac{1}{u} + \frac{1}{v} \rightarrow \frac{1}{0.50} = \frac{1}{0.52} + \frac{1}{v}$$

$$\frac{1}{v} = \frac{1}{0.50} - \frac{1}{0.52} \rightarrow \frac{1}{v} = \frac{2}{26}$$ so $v = 13$ m

Magnification $m = \frac{v}{u} \rightarrow m = \frac{13}{0.52}$ so $m = 25$

Now try this

1 A magnifying glass with a focal length of 8.0 cm produces a virtual image 24 cm from the lens. Use the lens formula to calculate the position of the object and hence find the magnification. **(2 marks)**

2 An object is placed at 0.20 m from a lens of focal length 0.10 m. Find the image position and calculate the magnification achieved. **(2 marks)**

Plane polarisation

Light is electromagnetic waves, and as a transverse wave it can be polarised.

Representing plane-polarised light

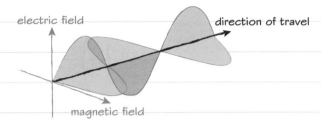

electric field

direction of travel

magnetic field

Light is a part of the electromagnetic spectrum. It travels as variations in the electric and magnetic fields in space at $c = 3.00 \times 10^8\,\text{m s}^{-1}$ in a vacuum. The field directions and the direction of travel are all mutually perpendicular.

In an **unpolarised** beam of light the field variations take place in all possible planes. **Plane-polarised** light has field variations in one plane only. Only the electric field variation is shown in these diagrams. In plane-polarised light the magnetic field variation also occurs in one plane only, at right angles to the electric field.

unpolarised light   plane-polarised light

Polarising filters

Polarised light occurs naturally: the light scattered towards the Earth by tiny particles in the upper atmosphere is partially plane polarised; light reflected from surfaces is also partially polarised – at a particular angle it is completely plane polarised. Certain types of crystal also polarise light passing through them.

Polarising filters polarise light. When unpolarised light falls onto the filter, it allows field variation through in only one plane so the emerging light is plane polarised.

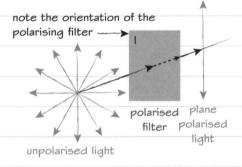

note the orientation of the polarising filter

polarised filter   plane polarised light

unpolarised light

Worked example

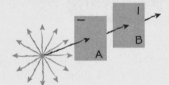

A B

A beam of unpolarised light is passed through polarising filter A as shown, then passes through a second polarising filter B. Filter B is the same as filter A, but rotated through 90°.

Label the diagram to show the nature of the light after passing through A and then through B. **(2 marks)**

Light is plane-polarised horizontally after filter A.

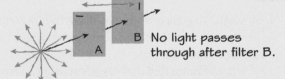

A   B   No light passes through after filter B.

Filter at 90° – the filters are said to be crossed.

Now try this

1. Polarising filters are used in sunglasses to remove the glare caused by light reflected from horizontal surfaces, like the surface of a swimming pool. Explain how this works. **(2 marks)**

2. Polarised light is also used to examine crystals and to show up regions of stress in engineering models of structures. The toughened glass used in car windscreens reveals such stress patterns under polarised light, and you may notice the patterns, when the Sun is low in the sky. Explain why this is so. **(4 marks)**

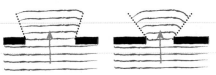
Diffraction and Huygens' construction

Diffraction is the spreading out of waves as they pass through gaps and around obstacles.

Observing diffraction

The ripple tank model shows that waves spread out as they pass through gaps, and that the amount of spreading increases as the gap size: wavelength ratio gets smaller.

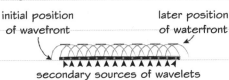

plane wave fronts travelling
towards a gap in a barrier

Huygens' construction

The Dutch scientist Huygens proposed a method for determining the position of a wavefront, given its initial position. He said every point on a wavefront could be considered as a source of **wavelets** that spread out in a circle at the same speed. The new position is found from the line that joins all the wavelets.

Huygen successfully used this method to explain observations of reflection at a surface and refraction as waves travelled from one material into another.

It also provided an explanation for the way light spread out on passing through a single slit, around small obstacles and through multiple slits.

initial position
of wavefront

later position
of waterfront

secondary sources of wavelets

Diffraction through a single slit

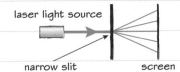

laser light source

narrow slit   screen

When a laser is directed onto a narrow slit the light is spread out and forms bright spots where the secondary wavelets are in phase and interfere constructively. These bright bands are separated by dark bands where destructive interference occurs.

Diffraction through double slits

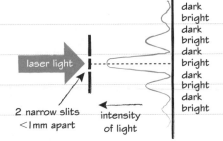

laser light

2 narrow slits
<1mm apart

intensity
of light

dark
bright
dark
bright
dark
bright
dark
bright
dark
bright

Here the double slit produces an interference pattern, with bright spots where light waves from each slit arrive at the screen in phase. A ripple tank model of two-source interference is described on page 42.

Diffraction grating

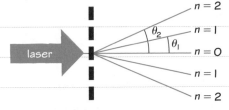

laser

$n = 2$
$n = 1$
θ_2
θ_1
$n = 0$
$n = 1$
$n = 2$

A **diffraction grating** consists of a slide on which very closely spaced vertical lines have been drawn.

When lit with **monochromatic** (single-frequency) light the grating produces bright spots or **maxima** at angles λ where the arriving waves are in phase.

$$n\lambda = d\sin\theta$$

where n is the order of the spectrum or spot and d is the separation between lines on the grating.

When lit with white light, the grating produces spectra with different wavelengths of light constructively interfering at different angles.

A green laser producing light of wavelength 520 nm is shone through a diffraction grating with 500 lines per mm.

(a) Calculate the angles at which the first- and second-order maxima (bright spots) are formed. **(4 marks)**

$n\lambda = d\sin\theta$,

$\lambda = 520 \times 10^{-9}\,\text{m}, \quad d = \dfrac{1}{500 \times 10^3}$

$\qquad\qquad = 2.00 \times 10^{-6}\,\text{m}$

For $n = 1$, $\theta = \sin^{-1}\left(\dfrac{520 \times 10^{-9}}{2.00 \times 10^{-6}}\right) = 15.1°$

For $n = 2$, $\theta = \sin^{-1}\left(\dfrac{2 \times 520 \times 10^{-9}\,\text{m}}{2.00 \times 10^{-6}}\right)$

$\qquad\qquad = 31.3°$

(b) What is the highest order maximum that can be produced? **(4 marks)**

$\sin\theta$ must be ≤ 1, so $\dfrac{2.00 \times 10^{-6}}{520 \times 10^{-9}} \leq 1$ and thus

$n \leq \dfrac{d}{\lambda} \rightarrow n \leq = 3.85$. As n must be an integer, the highest order is 3.

Using a diffraction grating to measure the wavelength of light

Apparatus for the investigation

Calculate d, the line spacing, in m. Project the diffracted beams onto a screen 1–3 m from the grating and measure this distance, D, using a metre rule or steel tape measure. Measure the distance, y, between a number of clear maxima – in the diagram, this is between the two second-order maxima. Making D, and therefore y, as large as practicable will reduce the percentage uncertainty in these values.

The line spacing will be given on the diffraction grating, usually in lines mm^{-1}.

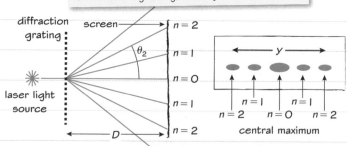

SAFETY ADVICE: Wear correct eye protection and do not look directly into the laser beam.

Worked example

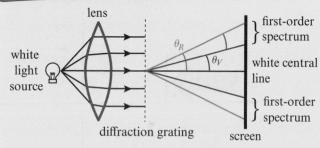

When white light from a tungsten filament lamp is passed through a diffraction grating a white central line is formed and spectra are formed on either side.

The lines on the grating are 1.5 μm apart.

(a) Calculate the wavelength of light at the red end of the first-order spectrum if θ_R is 27.82°. **(2 marks)**

$n = 1$ (first-order spectrum)

so $\lambda_R = d\sin\theta_R = 1.5 \times 10^{-6} \sin 27.82°$

Wavelength $\lambda_R = 700 \times 10^{-9}$ m

(b) Find θ_V if the wavelength of violet light is 400 nm. **(2 marks)**

$\theta_V = \sin^{-1}\left(\dfrac{\lambda_V}{d}\right) = \sin^{-1}\left(\dfrac{400 \times 10^{-9}}{1.5 \times 10^{-6}}\right)$

So violet light is diffracted through 15.47°.

Worked example

A diffraction grating with 200 lines mm^{-1} is placed 1.5 m from a screen and red laser light is shone through it. The distance between the two second-order maxima was found to be 0.81 m. Calculate the wavelength of light emitted by the laser. **(4 marks)**

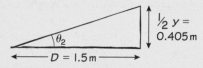

$n\lambda = d\sin\theta$ so $\lambda = \dfrac{d\sin\theta}{n}$

$n = 2$ (second order)

$d = \dfrac{1}{2.00 \times 10^5} = 5.00 \times 10^{-6}$ m

(200 lines mm^{-1} = 2.00 × 10^5 lines m^{-1})

$\theta_2 = \tan^{-1}\left(\dfrac{0.405}{1.5}\right) = 15.1°$

$\lambda = \dfrac{5.00 \times 10^{-6} \sin 15.1°}{2} = 651 \times 10^{-9}$ m

or 651 nm

Now try this

1 Light of wavelength $\lambda = 700$ nm is shone through a diffraction grating with 600 lines mm^{-1}. What is the highest-order maximum observed? **(4 marks)**

2 The diagram shows a diffraction grating illuminated with light from a laser. A third-order maximum is formed at an angle θ_3 from the normal. Find the distances BP and CQ:
(a) in terms of d and θ_3 **(2 marks)**
(b) in terms of λ, the wavelength of light from the laser. **(2 marks)**

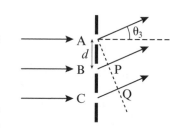

Electron diffraction

The discovery of electron diffraction showed that matter, like light, has wave properties.

Demonstrating electron diffraction

The strongest experimental evidence for matter behaving like waves comes from superposition effects.

These effects can be demonstrated for electrons by directing a beam of electrons at a thin slice of a polycrystalline material such as graphite. Each graphite crystal has a structure of parallel layers of atoms that are the right distance apart to act like the slits in a diffraction grating. The electron waves are diffracted, and constructive and destructive interference occurs at particular angles.

Many such crystals are arranged in all orientations in each piece of graphite, so the maxima form regularly spaced concentric rings on the screen.

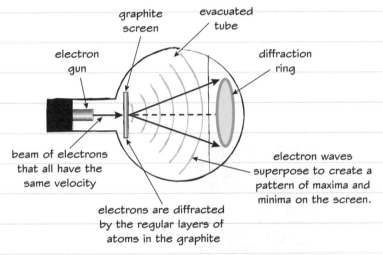

On page 54 some of the reasons to consider light as particles or waves are listed, along with de Broglie's ideas about electrons behaving in the same way as light.

The de Broglie equation

De Broglie's equation links the particle property of electrons, momentum $p \, (= mv)$, to their wave property, wavelength λ:

$$\lambda = \frac{h}{p}$$

where h is the Planck constant.

The faster the electrons are moving, the greater their momentum and the shorter their wavelength.

The equation applies to all matter, but the effects are only noticeable on the atomic scale.

Calculate the de Broglie wavelength of an electron travelling at 2% of the speed of light.
($c = 3.0 \times 10^8 \, \text{m s}^{-1}$, $h = 6.63 \times 10^{-34} \, \text{J s}$, $m_e = 9.1 \times 10^{-31} \, \text{kg}$) **(2 marks)**

$$\lambda = \frac{h}{mv} = \frac{6.63 \times 10^{-34}}{9.1 \times 10^{-31} \times 0.02 \times 3.0 \times 10^8}$$
$$= 1.2 \times 10^{-10} \, \text{m}.$$

This wavelength is much shorter than the shortest wavelength of visible light. High-speed electrons are used in electron microscopes because the shorter wavelength means that much smaller objects can be resolved than with visible light.

Now try this

1 Calculate the de Broglie wavelength of an electron that has been accelerated to $1.2 \times 10^5 \, \text{m s}^{-1}$.
($h = 6.63 \times 10^{-34} \, \text{J s}$; $m_e = 9.1 \times 10^{-31} \, \text{kg}$) **(2 marks)**

2 Explain why a proton accelerated through a p.d. V has a shorter de Broglie wavelength than an electron accelerated through the same p.d. **(4 marks)**

Waves and particles

No simple model is perfect to describe light, so the history of how the theory of light developed contains a very long argument about it.

Two theories are established

Scientists and philosophers have speculated about the nature of light for centuries. On the evidence of its behaviour, some suggested light was a stream of particles, others hypothesised that it was made of waves.

In the 17th century, scientists used both models to explain the behaviour of light, because both theories could explain reflection and refraction. Descartes and Huygens favoured the wave explanation whilst Newton was convinced that light was a stream of particles. At this time the stature of Newton led to wider acceptance of the particle theory.

Interference and Young's slits

Young demonstrated that light waves can superpose and cause interference patterns. This was strong evidence that light was a wave phenomenon; this and other experiments led to the particle theory being discarded.

Thomas Young
1773–1829

Photoelectric emission

The photoelectric effect studied by Einstein and the quantisation of energy discovered by Max Planck (1858–1947) were effects that could not be explained by the wave theory. Some effects required a different theory. The particle explanation was back on the scene.

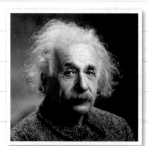

Albert Einstein
1879–1955

Particles behave like waves

Inspired by Einstein's work, de Broglie hypothesised that electrons might have wave-like properties. This was confirmed in 1927 by an experiment showing diffraction of electrons fired at a crystal surface.

The de Broglie equation and electron diffraction by a crystal are covered on page 53.

Louis de Broglie
1892–1987

So which theory wins?

The answer is neither. Waves and particles are both models that help us to understand the behaviour of matter. Centuries of scientific debate, in which scientists tested each others' experimental results and arguments, brought scientific consensus closer to a 'true picture'. Electromagnetic radiation is a stream of **photons** that have wave properties and discrete amounts of energy dependent on their frequency. Particles have an associated wavelength that can be determined by the de Broglie formula. Both facets of matter have been shown by experiment.

Worked example

Use the de Broglie formula to work out the wavelength associated with an electron travelling at 6000 km s^{-1}.
(mass of an electron $m_e = 9.11 \times 10^{-31}$ kg, Planck's constant $h = 6.63 \times 10^{-34}$ J s)   **(2 marks)**

$p = m_e v = 9.11 \times 10^{-31} \times 6.00 \times 10^6$

$\lambda = \dfrac{h}{p}$

$\lambda = \dfrac{6.63 \times 10^{-34}}{9.11 \times 10^{-31} \times 6.00 \times 10^6}$

$= 0.12$ nm (1.2×10^{-10} m)

Now try this

A cricket ball has a mass of 163 g and travels at 40 m s^{-1}. Use the de Broglie equation to calculate the associated wavelength of the cricket ball.
Comment on your result.   **(3 marks)**

The photoelectric effect

The explanation of the photoelectric effect marked birth of quantum physics.

An experiment with static electricity

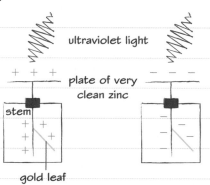

ultraviolet light

plate of very clean zinc

stem

gold leaf

Worked example

Explain why nothing happens when the zinc plate of a positively charged GLE is illuminated with UV light. **(2 marks)**

The zinc plate of the GLE loses electrons and so the system of plate, stem and leaf becomes more positively charged. But the leaf does not diverge more because electrons emitted are immediately attracted back to the positively charged plate.

One GLE is given a positive charge and the other a negative charge. When UV light is shone on each zinc plate, the gold leaf in the negatively charged GLE collapses, showing that it has lost charge. The positively charged GLE is unaffected.

Experimental observations

The emission of photoelectrons depends on:
- the wavelength, or frequency, of light shone on the metal
- the type of metal.

The intensity of the light has no effect on whether electrons are emitted.

Other experiments showed that:
- The maximum kinetic energy of emitted photoelectrons depends on the wavelength of light and the metal under test.
- Photoelectric emission is spontaneous – it occurs immediately light is shone onto metal, subject to the above conditions, no matter how low the light intensity is.

Quantum theory

The experimental evidence suggests that light and other EM radiation comes in 'packets' of energy called quanta. A quantum of EM energy is called a **photon**. The energy, E, of a photon depends on its associated frequency, f, and is given by:

$$E = hf$$

where h is Planck's constant.

Each photon will cause an electron to be emitted from a metal surface if the photon energy is large enough.

Photoelectric emission

The experiment shows that, under certain conditions, electrons are emitted from the surface of metals when illuminated with light. This process is called **photoelectric emission**.

Problems with the observations

The predictions of the wave theory of light failed to explain the photoelectric effect. It predicted that photoelectric emission would be dependent on the intensity of the light, not its wavelength. It takes energy to free electrons from the metal, and greater light intensity is associated with larger amplitude waves. The wave theory predicts that if low-intensity light were shone on the plate for enough time eventually electrons would acquire enough energy to escape – the effect would not be spontaneous.

The photoelectric equation

This is a conservation of energy equation:

photon energy	=	minimum energy needed to free an electron	+	maximum KE of emitted photoelectron

$$hf = \phi + \tfrac{1}{2}mv^2_{max}$$

ϕ is called the **work function** of the metal used. This is the minimum energy needed to free an electron from the metal surface. More energy is needed to free electrons below the surface layer.

If the photon energy $hf > \phi$ the energy surplus is transferred as KE to the electron emitted.

Now try this

1 For potassium $\phi = 3.65 \times 10^{-19}$ J. Find the maximum wavelength of light that will cause photoelectrons to be emitted when it is shone onto potassium. **(4 marks)**

2 State and explain how increasing the intensity of light of one wavelength directed at a metal surface will affect:
   (a) the number of photoelectrons emitted **(2 marks)**
   (b) their maximum kinetic energy. **(2 marks)**

Line spectra and the eV

Line spectra are the 'fingerprints' of elements.

Continuous and line spectra

White light produced by a tungsten filament lamp produces a **continuous spectrum**.

When elements in the gas or vapour state are heated until they emit light they do so at certain distinct wavelengths which are detected by the human eye as specific colours. These are **line spectra** and the position and number of lines are unique to each element.

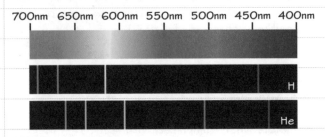

700nm 650nm 600nm 550nm 500nm 450nm 400nm

H

He

A continuous emission spectrum (top) and the emission line spectra of hydrogen (centre) and helium (bottom).

The electronvolt

It is usually more convenient to measure tiny amounts of energy in **electronvolts** than joules. 1 eV is the energy gained by an electron accelerated through a potential difference of 1 V.

$$1 \text{ eV} = 1.60 \times 10^{-19} \text{ J}$$

A problem

The model of the atom that describes electrons circling the nucleus does not match observations. Because the electrons are in a state of continuous acceleration, they should radiate energy and would eventually crash into the nucleus.

Bohr's solution

Bohr hypothesised that electrons in certain special orbits were stable. He calculated the energy values asssociated with these stable orbits for an atom of hydrogen. He did not have an explanation for this.

Allowed energy states

Bohr calculated the allowed energy levels for the electron in a hydrogen atom. The electron is normally in the lowest energy level (the 'ground state') but can be raised to a higher energy level if it absorbs the right amount of energy. It may only occupy specific energy states: to raise a ground-state electron to the first **excited** state (1→2) requires the specific energy {(−3.4) − (−13.6)} eV = 10.2 eV

0 eV ═══════

−0.85 eV ─────── 5 ⎤ excited
−1.51 eV ─────── 4 ⎥ states
        ─────── 3 ⎦
−3.4 eV ─────── 2

−13.6 eV ─────── 1

The permitted energy levels for the electron in the hydrogen atom.

What is the wavelength of the photon needed to raise the electron in an atom of hydrogen from level 1 to 2?   **(3 marks)**

$$10.2 \text{ eV} = 10.2 \times 1.60 \times 10^{-19} \text{ J}$$

$$E = hf \text{ and } f = \frac{c}{\lambda}$$

$$\lambda = \frac{hc}{E} = \frac{6.63 \times 10^{-34} \times 3 \times 10^{8}}{10.2 \times 1.6 \times 10^{-19}}$$

$$= 1.22 \times 10^{-7} \text{ m}$$

This is 122 nm and part of the UV spectrum.

Explaining emission spectra

An electron in an excited state will eventually drop back to the ground state, emitting a photon with a particular quantum of energy and therefore specific wavelength. It may drop to the ground state directly or in stages. In the diagram the emitted photon has resulted from a transition from level 3 to 2.

6 5
4
3
2   2

absorbed photon   emitted photon

1

Refer to the energy level diagram for hydrogen above.

1   Why do energy transitions between excited states and the ground state not produce lines in the visible part of the hydrogen emission spectrum?   **(3 marks)**

2   The visible emission spectrum of hydrogen contains a line at 434 nm that results from an electron in energy level 5 dropping back to level 2. Calculate the energy of level 5 in electronvolts.   **(4 marks)**

Exam skills 6

This exam-style question uses knowledge and skills you have already revised. Have a look at pages 52–56, for a reminder about the quantum nature of light.

Worked example

A red LED emits light of wavelength 630 nm. It is connected into a circuit like the one in the figure, and the supply voltage is slowly increased until the LED is just seen to glow.

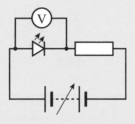

The reading on the voltmeter at this moment is 1.96 V. (1 eV = 1.60 × 10⁻¹⁹ J)

(a) Calculate the work done, in electronvolts and in joules, on a single electron as it moves through the LED. **(2 marks)**

One electronvolt is the energy transferred to one electron accelerated through one volt, so

$W = 1.96 \, eV$

$= 1.60 \times 10^{-19} \times 1.96 = 3.14 \times 10^{-19} \, J$

(b) Assume that all of the energy supplied to the electron as it moves through the LED is transferred to one photon of red light of wavelength 630 nm. Use this assumption to calculate a value for the Planck constant. **(3 marks)**

Photon energy $E = hf = \dfrac{hc}{\lambda}$.

If all of the work done on the electron is transferred to one photon, then
$E = 3.14 \times 10^{-19} \, J$.

Rearranging to make h the subject:

$h = \dfrac{E\lambda}{c} = \dfrac{3.14 \times 10^{-19} \times 630 \times 10^{-9}}{3.0 \times 10^{8}}$

$= 6.59 \times 10^{-34} \, Js$

(c) When the supply voltage is increased, the LED glows more brightly but the colour of the light it emits is unchanged. Explain this in terms of photons. **(4 marks)**

Increasing the supply voltage will increase the current flowing through the LED. More electrons per second will pass through the LED, so more photons per second will be emitted. This accounts for the increase in brightness.

The fact that the colour of the light does not change means that the wavelength is still the same, so the photons have the same energy as before. Therefore each electron is transferring the same amount of energy to a photon as it passes through the LED.

Worked example

Explain, with the aid of calculation, whether or not photoelectric emission can occur. **(3 marks)**

Photoelectric emission occurs only if the photon $hf \geq \phi$

$f = \dfrac{c}{\lambda}$, so $hf = \dfrac{hc}{\lambda}$

$= \dfrac{6.63 \times 10^{-34} \times 3.00 \times 10^{8}}{(1.87 \times 10^{-9})}$

$= 10.6 \times 10^{-19} \, J$

Yes, it can occur.

Command word: 'calculate'

When you are asked to calculate something you should always show your working.

 It is good practice to start with the algebra, rearranging equations if necessary, then substitute values and finally calculate your answer.

Don't forget to include the units! If you get stuck, you can work out the units from the equation.

Command word: 'explain'

If a question asks you to explain something, make sure you:

☑ Address all parts of the question: in part (c) there are **two** things to explain.

☑ Respond to any directions in the question: in part (c) you are asked to explain in **terms of photons**, so make sure that is what you do.

☑ Think first and write your answer in a clear logical sequence.

Impulse and change of momentum

If a force acts on something, the longer it acts the greater the change of momentum it produces.

Impulse

Provided there are no external forces acting, Newton's second law of motion can be expressed as an equation: $F = \dfrac{\Delta p}{\Delta t}$

If this is rearranged:

$F\Delta t = \Delta p = mv - mu$

Force × time ($F\Delta t$) is called **impulse**.

Impulse = change of linear momentum

The SI unit for impulse is the same as that of momentum: kg m s^{-1} or N s.

Impulse can also be calculated from the area under a graph of force against time.

Crumple zones

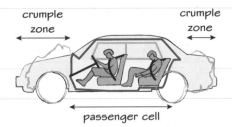

crumple zone   crumple zone

passenger cell

During a collision the passengers in a car undergo a rapid change of momentum. This means they experience a large impulse. Crumple zones increase the time in which they come to rest, and so reduce the force applied to the passengers.

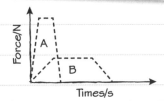

Force/N

A

B

Times/s

A: no crumple zones – short stopping time – large force on passengers
B: crumple zones – longer stopping time – smaller force on passengers
The areas of A and B are equal: the change of momentum is the same in both cases.

Using force–time graphs

The bobsleigh is accelerated by four athletes who then leap inside.

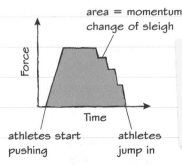

area = momentum change of sleigh

Force

Time

athletes start pushing   athletes jump in

If the applied force is plotted against time then the area under the graph is the impulse applied to the sleigh.

Worked example

A ball of mass 0.20 kg falls vertically and hits the ground at a velocity of $4.2\,\text{m s}^{-1}$. It rebounds vertically at an initial speed of $3.4\,\text{m s}^{-1}$.

(a) Calculate the change of momentum of the ball.   **(3 marks)**

Taking downward as positive:

$mu = 0.20 \times 4.2 = 0.84\,\text{kg m s}^{-1}$

$mv = -0.20 \times 3.4 = -0.68\,\text{kg m s}^{-1}$

$\Delta p = mv - mu = -0.68 - 0.84$
$\qquad = -1.52\,\text{kg m s}^{-1}$

(b) What is the impulse applied to the ball by the ground?   **(1 mark)**

Impulse = change of momentum
$\qquad = -1.52\,\text{N s}$ upwards

(c) The average force exerted on the ball while it was in contact with the ground was 18.0 N. Calculate the contact time.   **(2 marks)**

Impulse = $F\Delta t$ so $\Delta t = \dfrac{1.52}{18} = 0.084\,\text{s}$ (2 s.f.)

Now try this

A small model rocket-propelled car has a motor which burns for 2.0 s and provides an impulse of 5.0 N s. The car and its rocket motor have a combined mass of 100 g.

(a) What change of momentum can the motor provide for the car?   **(1 mark)**
(b) What is the average force applied to the car by the motor?   **(1 mark)**
(c) Calculate a value for the maximum velocity of the car based on the impulse provided by the motor.   **(2 marks)**
(d) Give two reasons why the actual maximum velocity of the car is likely to be different from the value you have calculated in (c).   **(2 marks)**

Conservation of momentum in two dimensions

Every component of linear momentum is conserved separately.

A two-dimensional collision

① Choose two perpendicular directions with the origin at the position of impact.

② Resolve the momentum along each direction.

③ Use conservation of momentum along each axis to solve problems.

 Maths skills This is a case of resolving vectors along two perpendicular axes.

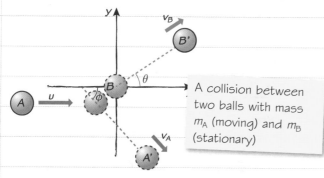

A collision between two balls with mass m_A (moving) and m_B (stationary)

In this diagram of a collision between two balls:

- In the x-direction:

$m_A u = m_A v_A \cos \theta + m_B v_B \cos \phi$

- In the y-direction:

$0 = m_A v_A \sin \theta - m_B v_B \sin \phi$

By selecting the x-axis parallel to the direction of motion of the incoming ball the problem is simplified. There is no initial y-component of momentum.

Using a vector triangle

Momentum is conserved, so the sum of momentum vectors before a collision must equal the sum after the collision. This can be drawn as a vector diagram:

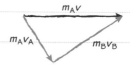

Worked example

A football of mass m is kicked against a wall. It approaches the wall at speed v and rebounds at the same speed as shown below.

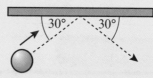

Use the law of conservation of momentum to explain why the ball rebounds at the same angle. **(3 marks)**

Each component of momentum is conserved. The component parallel to the surface of the wall is $mv \cos \theta$. m and v are unchanged by the collision so θ must be the same too.

Now try this

1 A car of mass 1400 kg is travelling at a speed of 25 m s^{-1} along a road on a bearing of 020°. This means 20° east of north.
   (a) Calculate the component of its linear momentum due north. **(2 marks)**
   (b) Calculate the component of its linear momentum due east. **(2 marks)**

2 If a collision is 'elastic' then kinetic energy is conserved as well as linear momentum.
   Look at the example of two balls colliding at the top of this page. Write down an equation for the conservation of kinetic energy in this collision. **(3 marks)**

Elastic and inelastic collisions

All collisions and interactions conserve linear momentum, but very few also conserve kinetic energy.

Collisions and conservation

Type	Kinetic energy	Momentum
elastic	conserved	conserved
inelastic	not conserved	conserved

Total energy is always conserved, but in an **inelastic collision** some of the initial kinetic energy (KE) is transferred to other forms, usually heat.

Linking momentum and kinetic energy

Momentum $p = mv$

Kinetic energy $E_K = \frac{1}{2}mv^2$

$p^2 = m^2v^2$ so $\dfrac{p^2}{2m} = \dfrac{m^2v^2}{2m} = \frac{1}{2}mv^2$

$E_K = \frac{1}{2}mv^2 = \dfrac{p^2}{2m}$

This is a very useful equation for switching between momentum and kinetic energy, especially in particle physics.

An inelastic collision

When two cars collide head-on, a great deal of work is done against internal forces, mainly in the plastic deformation of metal. This energy is eventually transferred as heat. The cars have very little kinetic energy after the collision.

Analysing a collision in one dimension

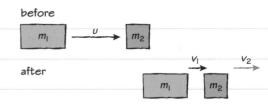

before

after

To analyse the collision you need to consider momentum and KE before and after the collision:

$m_1u = m_1v_1 + m_2v_2$ (conservation of momentum)

$\frac{1}{2}m_1u^2 = \frac{1}{2}m_1v_1^2 + \frac{1}{2}m_2v_2^2$ (elastic collision)

or

$\frac{1}{2}m_1u^2 > \frac{1}{2}m_1v_1^2 + \frac{1}{2}m_2v_2^2$ (inelastic collision)

A special case

In some cases the two bodies stick together after the collision ($v_1 = v_2$). This makes the calculations much easier. This is a special example of an inelastic collision.

Worked example

The diagrams show two objects before and after a collision.

before

5.0 kg $4.0\,\text{m s}^{-1}$ 2.0 kg

$2.0\,\text{m s}^{-1}$ v_2

after

(a) Calculate the final velocity of the 2.0 kg mass. **(2 marks)**

Momentum is conserved:

$m_1u = m_1v_1 + m_2v_2$

Rearranging: $v_2 = \dfrac{(m_1u - m_1v_1)}{m_2}$

$v_2 = \dfrac{5.0 \times 4.0 - 5.0 \times 2.0}{2.0} = 5.0\,\text{m s}^{-1}$

(b) Determine whether or not this collision is elastic. **(3 marks)**

If the collision is elastic, kinetic energy will be conserved.

$E_{K\,\text{before}} = \frac{1}{2}m_1u^2 = \frac{1}{2} \times 5.0 \times 4.0^2 = 40\,\text{J}$

$E_{K\,\text{after}} = \frac{1}{2}m_1v_1^2 + \frac{1}{2}m_2v_2^2$

$= \frac{1}{2} \times 5.0 \times 2.0^2 + \frac{1}{2} \times 2.0 \times 5.0^2 = 35\,\text{J}$

$E_{K\,\text{after}} < E_{K\,\text{before}}$ so some KE has been transferred to heat and sound (5.0 J). This is an inelastic collision.

Now try this

1 Two rugby players running in opposite directions collide and come to rest. Explain how such a collision (a) can conserve linear momentum but (b) does not conserve kinetic energy. **(4 marks)**

2 A car of mass 1200 kg travelling at $10\,\text{m s}^{-1}$ collides with a stationary truck of mass 3600 kg and the two vehicles lock together and move forwards.
(a) Calculate the velocity of the two vehicles just after the collision. **(3 marks)**
(b) Show that this is an inelastic collision and calculate the energy transferred to other forms in the collision. **(4 marks)**

Investigating momentum change

Forces and momentum change can be measured and related experimentally.

 Practical skills

The relationship between the force exerted on an object and its change of momentum

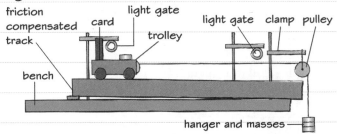

The equipment is set up so that the mass hanger will hit the floor just as the trolley passes the second light gate. The only horizontal force on the trolley is the tension in the string. The trolley starts from rest just before the first light gate and time t is the measured time for the trolley to move from the first to the second light gate. Masses are moved from the hanger, where they provide the force accelerating the trolley onto the trolley. The journey during which the force is applied is timed. The final velocity v with which the trolley passes the second light gate is recorded, from the time for which the known width of card eclipses the light gate. Data are obtained for a graph showing how different forces affect the change on momentum of the same mass.

Force mg acts on the trolley for time of travel t, where m is the hanging mass. Momentum increases from 0 to Mv where v is the final velocity and M is the total mass of trolley, string and mass hanger.

force × time = change of momentum

Force × time is called **impulse** and has units of N s.

Impulse $Ft = mgt = Mv$

A plot of mt against v should thus yield a straight line with gradient $\frac{M}{g}$.

Refer back to page 58 to see more about change in momentum.

Analysing collisions between small spheres

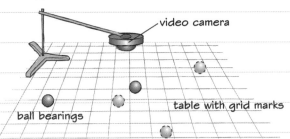

Two-dimensional collisions can be investigated by analysing video footage frame by frame. In this case, ball-bearings colliding on a table covered with a measured grid can be filmed from above. With measurement scales in two dimensions, the components of velocity in each dimension can be isolated. A spreadsheet filled out with the resulting measurements can calculate changes in velocity. Thus, separate calculations can be made in each dimension, in order to verify the conservation of momentum in two dimensions.

Now try this

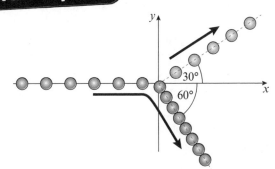

The diagram shows images of a collision between two balls each of mass 0.30 kg. The images show the positions of the two balls at regular intervals. The red ball is initially moving at 5.0 ms⁻¹ to the right and the blue ball is initially at rest.

(a) How can you tell from the image that the velocity of the incoming ball is constant. Explain your answer. **(2 marks)**

(b) How can you tell from the image that the speed of the red ball is greater than the speed of the blue ball after the collision? **(1 mark)**

(c) Calculate the speeds of the two balls after collision. **(5 marks)**

Exam skills 7

This exam-style question uses knowledge and skills you have already revised. Have a look at pages 14, 58–61, for a reminder about impulse, momentum and elastic and inelastic collisions.

Worked example

A man pushes a child on a sledge along horizontal snow with a force that varies in the way shown in the figure. The sledge is initially at rest. The man stops pushing after 3.5 s and lets it slide away. The frictional drag from the snow is negligible. The total mass of the sledge and child is 75 kg.

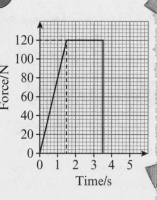

(a) State the physical quantity represented by the area under the graph and give its correct SI unit. **(1 mark)**

Impulse (or change of momentum).
SI unit is N s or kg m s^{-1}.

(b) Calculate the velocity of the sledge at the moment the man stops pushing. **(4 marks)**

impulse = change of momentum
       = area under graph.
area = $\frac{1}{2}$ × 120 × 1.5 + 120 × 2.0 = 330 N s
$m\Delta v$ = 330 so $\Delta v = \frac{330}{75}$ = 4.4 m s^{-1}
Since u = 0, the final velocity is 4.4 m s^{-1}.

(c) Later, when the same child and sledge is sliding over horizontal snow at 4.0 m s^{-1}, a second child of mass 45 kg runs in the same direction at 6.0 m s^{-1} and jumps onto the moving sledge.

(i) Calculate the velocity of the sledge with two children just after the second child lands on it. **(2 marks)**

momentum is conserved so:
75 × 4.0 + 45 × 6.0 = 570 = (75 + 45)v
$v = \frac{570}{120}$ = 4.75 = 4.8 m s^{-1} to 2 s.f.

(ii) State whether this is an elastic or inelastic collision and justify your answer by calculation. **(3 marks)**

KE before = $\frac{1}{2}$ × 75 × 4.0^2 + $\frac{1}{2}$ × 45 × 6.0^2
         = 1410 J
KE after = $\frac{1}{2}$ × 120 × 4.75^2 = 1354 J

There is a small difference in KE before and after. It is not conserved in the collision so the collision is inelastic.

If a question says that something is initially 'at rest' this is telling you that u = 0.

Command word: State

If a question asks you to state something, just do that! In part (a), simply writing 'impulse' is sufficient.

The quantity represented by the area under a graph is the product of the quantities on the two axes, in this case force × time or impulse.

You might know that the unit of impulse is N s, but if not you can work it out by multiplying the units from the two axes: N × s.

The easiest way to calculate the area is to divide it into a triangle, up to 1.5 s, and a rectangle, from 1.5 s to 3.5 s. The area of a triangle is half the base times the height.

To answer this part you need to use the mass of the child and sledge given at the start of the question (75 kg).

Your answer will be easier to understand if you start by stating the physical principle you are using – in this case the law of conservation of momentum.

There are two parts to this question – the statement and the justification. Make sure you do both!

Describing rotational motion

Before you can analyse rotation you need to understand the key terms that are used to describe this kind of motion.

Degrees and radians

The diagram shows an angle θ drawn inside a circle of radius r. The length of the arc opposite the angle is l.

One **radian** is the angle for which the arc length is equal to the radius.

The angle in radians $= \dfrac{\text{arc length}}{\text{radius}}$

$\theta = \dfrac{l}{r}$

For a complete rotation, $l = 2\pi r$, so $\theta = 2\pi$

To convert between radians and degrees:

2π radians $= 360°$

1 radian $= \left(\dfrac{180}{\pi}\right)^{\circ} \approx 57°$

1 degree $= \dfrac{\pi}{180}$ radians

Maths skills If you are working out trigonometric functions such as sine or cosine using your calculator, make sure you switch it to the same units as the angles ('deg' or 'rad').

Time period of rotation

If the rotation period (time to complete one rotation) is T then the angular velocity is:

$\omega = \dfrac{2\pi}{T}$ or $T = \dfrac{2\pi}{\omega}$

Angular velocity

Angular velocity ω is the angle swept out by an object with rotational motion in unit time, so the SI unit for angular velocity is rad s^{-1}.

If the angular velocity is constant:

$\omega = \dfrac{\Delta\theta}{\Delta t}$

where $\Delta\theta =$ angle turned through

and $\Delta t =$ time taken

Maths skills The symbol 'Δ' is used to represent a change in some variable.

Tangential velocity and angular velocity

The object below is moving in a circle with constant speed v.

In time Δt the object moves a distance $l = v\Delta t$

The angle turned through is

$\Delta\theta = \dfrac{v\Delta t}{r}$ radians

Rearranging: $\dfrac{\Delta\theta}{\Delta t} = \dfrac{v}{r}$

This is a very useful equation:

$\omega = \dfrac{v}{r}$ or $v = \omega r$

Useful equations

$\theta = \dfrac{l}{r}$ $\omega = \dfrac{\Delta\theta}{\Delta t}$

$v = \omega r$ $T = \dfrac{2\pi}{\omega}$

Worked example

The Earth has radius 6400 km and rotates once on its axis in 24 hours.

(a) Calculate the angular velocity of the Earth. **(2 marks)**

The Earth turns through 2π radians in 24 hours so:

$\omega = \dfrac{2\pi}{24 \times 3600} = 7.3 \times 10^{-5}$ rad s^{-1}

(b) Calculate the tangential velocity of a person standing at the equator. **(1 mark)**

$v = \omega r = 7.3 \times 10^{-5} \times 6400 \times 10^3$

$= 470$ m s^{-1}

Now try this

1. An ice skater spins 3.0 times per second as part of an ice dance.
   (a) Calculate her angular velocity. **(2 marks)**
   (b) Her outstretched arms move in a circle of radius 0.90 m. What is the speed of her fingertips? **(2 marks)**

2. Old vinyl records were played at 45 rpm (revolutions per minute). Calculate the angular velocity of one of these records when it is playing. **(3 marks)**

3. London is at a latitude of 51° N. Use data from the worked example to answer this question.
   (a) Calculate the angular velocity of a person in London as the Earth rotates on its axis. **(1 mark)**
   (b) Calculate the tangential velocity of a person in London. **(2 marks)**

Don't forget to change the time in hours to seconds!

Uniform circular motion

An object moving in a circle is always accelerating toward the centre of the circle.

Changing velocity

An object in uniform circular motion has constant angular velocity and speed. However, its velocity is continually changing. This is because velocity is a vector and the direction of the object changes continuously.

The vector diagram shows the change in velocity Δv as the Moon moves from A to B. If it takes time Δt to move from A to B then there is an average acceleration:

$$a = \frac{\Delta v}{\Delta t}$$

This shows that an object in circular motion is **always accelerating**.

To find out the direction of the acceleration imagine that the time Δt is very short so that A and B are close together. The direction of Δv would be at 90° to the velocity, pointing in toward the centre of the circle. This is called a **centripetal acceleration**. Centripetal means 'centre-seeking'.

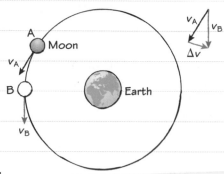

The Moon moves in approximately uniform circular motion as it orbits the Earth. Its speed is the same at A and B, but its direction is different.

🖩 **Maths skills** To understand the argument here you need to recall that vectors have both magnitude and direction. Velocity is a vector, and even though its magnitude, speed, is constant, its direction is changing – so the velocity is changing and the object is accelerating.

- -

Equation for centripetal acceleration

Considering positions A and, after a time t, B, the components of the object's velocity in the x and y directions are:

vertical: $v_y = v\cos\theta$ (the same in both cases, so the vertical acceleration is zero)

horizontal:

at A: $v_x = v\sin\theta$ at B: $v_x = -v\sin\theta$

So the acceleration a between A and B is just horizontal, and is:

$$a_x = \frac{2v(\sin\theta)}{t}$$

Since angular velocity $\omega = \frac{\theta}{t}$ and $v = r\omega$,

$$v = \frac{r\theta}{t}, \text{ so } t = \frac{r\theta}{v}$$

Here, the total angle moved in time t is 2θ, so:

$$t = \frac{r2\theta}{v}$$

$$\text{so } a_x = \frac{2v^2(\sin\theta)}{r2\theta}$$

To find the instantaneous acceleration at any point on the circumference, we reduce θ:

In the limit, as θ tends to zero, $\frac{\sin\theta}{\theta}$ tends to one:

$$a_x = \frac{v^2}{r}$$

But since $v = r\omega$,

$$a = \frac{(r\omega)^2}{r}$$

centripetal acceleration $a = r\omega^2$

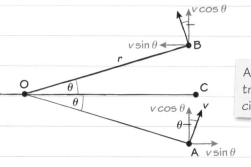

An object travelling on a circular path

Now try this

1 The Moon orbits the Earth every 27.3 days and the radius of its orbit is 380 000 km. Calculate the Moon's centripetal acceleration and state its direction. **(4 marks)**

2 A car is moving along a straight horizontal road at constant speed v. The diagram below shows one of its wheels and tyres. A and B are particles of dirt on the outside edge of the tyre.

State the magnitude and direction of the velocity of each particle of dirt relative to the road. **(3 marks)**

Centripetal force and acceleration

If something is accelerating, there must be a resultant force acting on it.

Centripetal force

From Newton's second law of motion, we know that if something is accelerating there must be a resultant force acting on it, and this force must act in the direction of the acceleration.

Objects moving in circular motion experience centripetal acceleration and thus require a resultant force acting toward the centre of the circle: a centripetal force. To review Newton's second law, see page 11

Centripetal acceleration:

$$a = \frac{v^2}{r} = r\omega^2$$

Centripetal force:

$$F = ma = m\frac{v^2}{r} = mr\omega^2$$

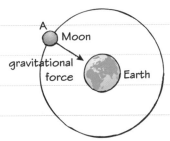

Gravitational attraction to the Earth provides the centripetal force that makes the Moon move in its circular orbit. This force always points toward the centre of the circle.

Looping the loop

In the diagram, an aircraft is moving in a circular path at constant speed. Centripetal force is provided by a combination of the aircraft's weight W, thrust T and lift L from the air.

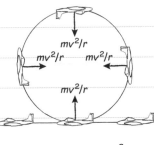

The **resultant force** at each point is $F = \frac{mv^2}{r}$ toward the centre of the circle.

For example, the vertical forces acting at the bottom of the loop:

Centripetal force
$$= \frac{mv^2}{r} = L - W$$

The diagram shows a track for a toy car. If the car is released from point A or above on the slope it will complete the loop without falling off.

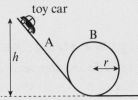

(a) The car has mass m and is released from the top of the slope. It easily completes the loop, passing the top point B with speed v. Draw a force diagram to show the forces acting on the car as it passes B. **(2 marks)**

Two forces act on the car, weight W and a normal contact force N from the track.

(b) Write down two expressions for the magnitude of the resultant force on the car at point B. **(2 marks)**

Centripetal force
$$F = W + N$$
$$F = \frac{mv^2}{r}$$

(c) Use your answers to (a) and (b) to find an expression for the contact force N at point B in terms of the gravitational field strength g. **(3 marks)**

$$N + W = \frac{mv^2}{r}$$
$$N = \frac{mv^2}{r} - W = \frac{mv^2}{r} - mg$$
$$N = m\left(\frac{v^2}{r} - g\right)$$

Now try this

1 Look at the worked example above.
   (a) Explain why below a certain minimum speed, the car will fail to complete the loop. **(4 marks)**
   (b) Derive an expression for the speed at which the car just fails to complete the loop. **(3 marks)**
   (c) Derive an expression for the minimum height, h, from which the car must be released in order for it to complete the loop. **(4 marks)**

2 The angular velocity of a DVD changes from 160 rad s⁻¹ when the player is reading near the centre to 70 rad s⁻¹ when it is reading near the edge. A particle of dust of mass 0.0050 mg (1 mg = 10^{-6} kg) is stuck to the disc at a distance of 2.0 cm from the centre of rotation. Calculate the minimum force that would just prevent it moving when the disc is played. **(3 marks)**

Electric field strength

Fields – a key concept in physics – allow objects that are separate in space to influence one another.

Forces and fields

Like charges repel, opposite charges attract.

These forces between charges can be explained by the idea of an **electric field**. Electric fields are defined as regions of space in which charged particles experience a force.

Defining the field strength

Electric field strength E is the force per unit charge acting on a positive test charge at a point in space.

$$E = \frac{F}{Q}$$

The S.I. units are $N\,C^{-1}$ (equivalent to $V\,m^{-1}$).

Forces are vectors, so electric fields are vector fields. This means that the field strength has a magnitude and a direction at every point in space. The direction is the direction in which force would be exerted on a positively charged particle.

Representing the electric field

Electric fields can be drawn as field lines.

- The direction of the line is the direction of the force on a positive charge.
- The closeness of the lines shows the strength of the field.

✓ Field lines start on positive charges.

✓ Field lines end on negative charges.

✓ Field lines cannot cross.

Radial fields

Radial field due to a point positive charge

Radial field due to a point negative charge

Uniform fields

Uniform fields have the same magnitude of field strength and the same direction at each point in space.

A uniform electric field can be set up between two charged plates.

The field lines are:

✓ parallel

✓ equally spaced.

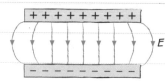

The field between the plates is uniform. It varies a little at the ends.

The field strength between two charged plates is:

$$E = \frac{V}{d}$$

where V is the p.d. between the plates and d is the distance between them.

Worked example

Sketch the field pattern formed when a point positive charge interacts with a point negative charge. **(2 marks)**

The resultant electric field strength is the vector sum of the field strengths at each point.

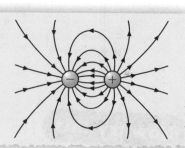

Now try this

1 Use the idea of an electric field to explain why an alpha particle, which is positively charged, would be repelled from a gold nucleus. **(3 marks)**

2 An oil droplet carrying a charge of 3.2×10^{-19} C is sprayed between two parallel metal plates between which there is a p.d. of 100 V and a distance of 5.0 cm.
(a) Calculate the electric field strength between the plates. **(2 marks)**
(b) Calculate the force on the oil droplet. **(2 marks)**

Electric field and electric potential

Electric field strength and electric potential are closely related, and both can be used to describe a field.

Forces and energy transfers in an electric field

Imagine pushing the positive charge Q from A (−) to B (+) in the uniform electric field E. You will have to apply a force $F = EQ$ in the opposite direction to the electrostatic force on the charge from the field. As you moved the charge, you would be doing work on it, and increasing its electric potential energy. This energy depends on the change in the **electric potential** and on the charge.

Electric potential at a point in the field is defined as energy per unit charge at that point.

A stronger electric field means that the potential changes more rapidly with distance as a charge moves through the field. There is a greater **potential gradient**. The negative potential gradient is equal to electric field strength, $E = -\dfrac{V}{d}$.

Electric potential

Electric potential $V = \dfrac{\text{electric potential energy}}{\text{charge}}$

$= \dfrac{EPE}{Q}$

S.I. unit for electric potential is $J\,C^{-1}$ or V.

Electric potential difference or just 'p.d.' is often referred to as the 'voltage' between two points in an electric circuit.

Equipotentials

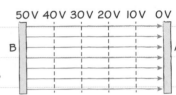

No work is done when a charge moves perpendicular to the field lines. This means that the potential is constant along lines perpendicular to the field lines. These lines are called **equipotentials**.

In a uniform field, the equipotentials are equally spaced.

Defining field strength E and potential V

$E = \dfrac{F}{Q}$ (force per unit charge) $V = \dfrac{EPE}{Q}$ (electric potential energy per unit charge)

Dividing by charge makes E and V **properties of the field** at a point in space.

Worked example

An electron is accelerated through a potential difference of 500 V between two electrodes 10.0 cm apart. The charge and mass of an electron are: $e = 1.60 \times 10^{-19}$ C and $m = 9.11 \times 10^{-31}$ kg.

(a) Calculate the electric field strength between the plates and state its direction. **(1 mark)**

$E = \dfrac{V}{d} = \dfrac{500}{0.10} = 5000\ V\,m^{-1}$, acting from the negative electrode toward the positive electrode.

(b) Calculate the force on the electron. **(1 mark)**

$F = eE = 1.60 \times 10^{-19} \times 5000 = 8.00 \times 10^{-16}$ N

(c) Calculate the velocity of the electron when it reaches the positive terminal. **(3 marks)**

$\frac{1}{2}mv^2 = QV$ so $v = \sqrt{\dfrac{2QV}{m}} = \sqrt{\dfrac{2 \times 1.60 \times 10^{-19} \times 500}{9.11 \times 10^{-31}}} = 13.3 \times 10^6\ m\,s^{-1}$

Now try this

A singly charged positive ion is accelerated through a potential difference of 350 V by two electrodes placed 8.0 cm apart. The ion has mass 3.8×10^{-26} kg.

(a) Calculate the electric field strength between the electrodes. **(2 marks)**

(b) Calculate the acceleration of the ion. **(2 marks)**

(c) Calculate the final velocity of the ion. **(2 marks)**

Forces between charges

Coulomb worked out how the force between two point charges varies with distance.

Coulomb's law

Radial electric fields, for a point charge, or a uniformly charged sphere, spread out evenly in all directions. As you get farther away from the charge, the field strength is smaller and the field lines are spread farther apart. The force between two charges gets weaker with distance.

Coulomb showed that the force between two small, point-like, charges Q_1 and Q_2 depends on:

 the product of the charges, Q_1Q_2

 the inverse-square of their separation.

$$F \propto \frac{Q_1Q_2}{r^2}$$

The constant of proportionality is $\frac{1}{4}\pi\varepsilon_0$, where ε_0 is the **permittivity of free space**

$\varepsilon_0 = 8.85 \times 10^{-12}\ \text{F m}^{-1}$

The full equation for **Coulomb's law** is thus:

$$F = \frac{Q_1Q_2}{4\pi\varepsilon_0 r^2}$$

Radial fields around point charges are described on page 69 and the unit F (for Farad) is explained on page 70 in the context of capacitors.

Forces between more than two charges

 Calculate the force between each pair of charges.

 Add the force vectors to find the resultant force.

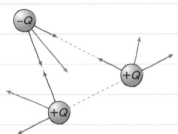

Worked example

The radius of a hydrogen atom is $5.3 \times 10^{-11}\ \text{m}$. Calculate the force exerted by the nucleus, charge $+e$, on the orbiting electron, charge $-e$. **(3 marks)**

From Coulomb's law,

$$F = \frac{e^2}{4\pi\varepsilon_0 r^2}$$

$$= \frac{(1.6 \times 10^{-19})^2}{4\pi \times 8.85 \times 10^{-12} \times [5.3 \times 10^{-11}]^2}$$

$$= 8.2 \times 10^{-8}\ \text{N}$$

Now try this

1 Two small insulated spheres 2.0 cm apart carry charges of $+20\,\text{nC}$ and $-50\,\text{nC}$, respectively.
   (a) Calculate the force on the 20 nC charge and state its direction. **(3 marks)**
   (b) Calculate the force on the 50 nC charge and state its direction. **(2 marks)**

2 In Rutherford's scattering experiment, alpha particles were fired at a thin gold foil. Most of them passed straight through, but a very few were deflected or even bounced back, showing that there must be very small, concentrated areas of positive charge within the gold atoms. Calculate the force on an alpha particle ($Q = +2e$) approaching a gold nucleus ($Q = +79e$) at a distance of (1 pm = 10^{-12} m).
   ($e = 1.60 \times 10^{-19}\,\text{C}$, $\varepsilon_0 = 8.85 \times 10^{-12}\,\text{F m}^{-1}$) **(3 marks)**

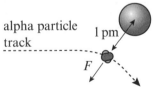
gold nucleus
alpha particle track
1 pm
F

Field and potential for a point charge

Point charges and uniformly charged spheres create radial fields.

Electric field around a point charge

Imagine placing a small positive charge Q_1 at a distance r from a point charge Q_2.

By Coulomb's law the force F on Q_1 will be:

$$F = \frac{Q_1 Q_2}{4\pi\varepsilon_0 r^2}$$

Since electric field strength $E = \frac{F}{Q_1}$,

$$E = \frac{Q_2}{4\pi\varepsilon_0 r^2}$$

or, generally,

$$E = \frac{Q}{4\pi\varepsilon_0 r^2}$$

E is a vector. The direction is the direction of force on a positive charge.

Electric field strength E is introduced on page 66.

Electric potential around a point charge

The potential at distance r from a point charge of magnitude Q is given by the formula:

$$V = \frac{Q}{4\pi\varepsilon_0 r}$$

Electric potential is a scalar. Its sign depends on the sign of the charge that causes the field.

Electric potential V is introduced on page 67.

Field E and potential V for a point charge:

$$E = \frac{Q}{4\pi\varepsilon_0 r^2}$$

$$V = \frac{Q}{4\pi\varepsilon_0 r}$$

Field lines and equipotentials in a radial field

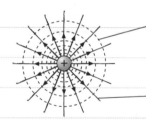

Equipotentials are perpendicular to field lines. Their separation increases as we move farther from the charge because $V \propto \frac{1}{r}$. For a positive charge the potential increases as you approach the charge (because work would have to be done to push a positive charge toward it).

Field lines are perpendicular to equipotentials. The field is strongest where they are closest together. They spread out according to an inverse-square law: $E \propto \frac{1}{r^2}$

Worked example

Sketch the resultant electric field and electric potential between a positive and a negative point charge on the following diagram using field lines and equipotentials. **(4 marks)**

⊕   ⊖

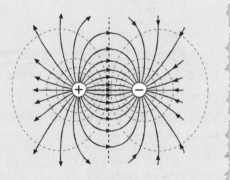

To find the field at a point in space:
- Find the field at that point caused by each individual charge.
- Add the individual field strengths, taking into account their directions.

Now try this

1 Calculate the electric field strength and electric potential at a distance of 25 mm from a point charge of size +50 nC. ($\varepsilon_0 = 8.85 \times 10^{-12}$ F m^{-1}) **(4 marks)**

2 Look at the diagram of the dipole field in the worked example above. The two charges are $+1.60 \times 10^{-19}$ C and -1.60×10^{-19} C, and the distance between them is 2.5×10^{-10} m. Calculate the field strength and potential at the mid-point between the two charges. **(4 marks)**

Capacitance

Capacitors are electrical components used to store charge and energy.

What is a capacitor?

A **capacitor** consists of two conductors separated by a **dielectric** (an insulator). It stores charge. The symbol for a capacitor is: ⊣⊢

Different types of capacitor.

Electrolytic capacitors (like the one on the left) must be connected with the correct polarity, otherwise they can explode! Notice that they all have two terminals – these connect to the two plates inside the capacitor.

Definition of capacitance

The charge stored by a capacitor is directly proportional to the final p.d. across the capacitor.

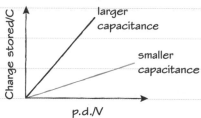

Capacitance is the charge stored per unit p.d.:

$$C = \frac{Q}{V} = \frac{Q}{V}$$

Beware! C (usually italic) is used for capacitance but C is also used for the S.I. unit of charge, the coulomb.

The S.I. unit for capacitance is the farad, F.

$1 F = 1 C V^{-1}$

Most capacitors have a capacitance measured in μF or nF.

When the capacitor is fully charged, the p.d. across it will be 12.0 V.

Storing charge

The circuit shows how a capacitor charges. Moving coil meters are better for this than digital meters.

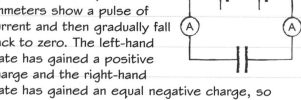

When S is closed, both ammeters show a pulse of current and then gradually fall back to zero. The left-hand plate has gained a positive charge and the right-hand plate has gained an equal negative charge, so there is now a p.d. across the capacitor equal but opposite to that of the battery. Opening and closing the switch again results in no further current – the capacitor is charged.

Discharging a capacitor

A charged capacitor has a p.d. across its plates, so if it is connected into a circuit charge will flow. The capacitor discharges until eventually there is no excess charge on either plate and the p.d. across it is zero.

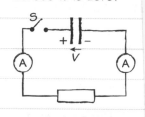

When switch S is closed both ammeters indicate a flow of charge around the circuit. This current decays to zero. Increasing the resistance reduces the current so the decay takes longer.

Worked example

A 220 μF capacitor is charged by a 12.0 V supply.

(a) Calculate the final charge on the capacitor. **(2 marks)**

$C = \frac{Q}{V}$

so $Q = CV = 220 \times 10^{-6} \times 12.0$

$= 2.64 \times 10^{-3} C = 2.64 \, mC$

(b) The charged capacitor is then discharged through a 100 Ω resistor. Calculate the initial discharge current. **(2 marks)**

$I = \frac{V}{R} = \frac{12.0}{100} = 0.12 A = 120 \, mA$

Now try this

The diagram shows the charge stored for different p.d.s across a capacitor.

(a) Calculate the capacitance of the capacitor. **(3 marks)**

(b) Calculate the charge stored on the capacitor when it is connected to a 12 V supply. **(2 marks)**

(c) The charged capacitor is then discharged through a 50 Ω resistor. Calculate the initial discharge current. **(2 marks)**

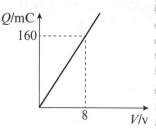

Energy stored by a capacitor

Capacitors can be used to store and release energy, for example in a camera flash gun.

Storing energy on the capacitor

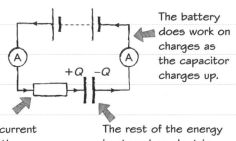

The battery does work on charges as the capacitor charges up.

Some energy is lost as heat as current passes through the resistor.

The rest of the energy is stored as electric potential energy on the charged capacitor.

Releasing energy from the capacitor

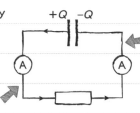

All of the energy stored on the capacitor is dissipated as heat as the discharge current passes through the resistor.

The energy stored on the capacitor decreases as the capacitor discharges.

Energy equations

As charge is stored on the capacitor, the energy stored is equal to the area under the graph of charge against p.d.

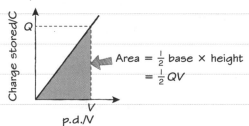

Area = $\frac{1}{2}$ base × height = $\frac{1}{2}QV$

Energy stored on a capacitor:

$$W = \frac{1}{2}QV$$

We can substitute from $C = \frac{Q}{V}$ to find two other forms of this equation:

$$W = \frac{1}{2}QV = \frac{1}{2}CV^2 = \frac{1}{2}\frac{Q^2}{C}$$

Worked example

Look at the charging circuit.

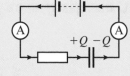

(a) Explain why the charging process is not 100% efficient. **(2 marks)**

Efficiency is the ratio of useful energy output to total energy input. In this case the useful output is energy stored on the capacitor. Some of the energy supplied by the battery is dissipated as heat in the resistor so the energy stored is less than the energy supplied and the efficiency is less than 100%.

(b) Show that the charging efficiency is actually 50%. **(2 marks)**

The work done by the battery when charge Q passes through it is $W_b = QV$. The energy stored on the capacitor is $W_c = \frac{1}{2}QV$ so the efficiency is $\left(\frac{W_b}{W_c}\right) \times 100\% = 50\%$

(c) Calculate the energy stored on a 50 µF capacitor when it is charged by a battery of e.m.f. 16 V. **(2 marks)**

$$W = \frac{1}{2}CV^2 = \frac{1}{2} \times 50 \times 10^{-6} \times 16^2$$
$$= 6.4 \times 10^{-3}\,\text{J}$$

Now try this

1　Look at the worked example about charging a capacitor. What effect would changing the circuit resistance have on the charging efficiency? **(1 mark)**

2　A 100 mF capacitor has a charge of +0.60 mC on its positive plate and then it is completely discharged through a resistor in a time of 10 ms.

(a) Calculate the energy stored on the capacitor. **(2 marks)**

(b) Calculate the average power supplied by the capacitor as it discharges. **(2 marks)**

(c) In (b) you calculated the average power supplied by the capacitor. Explain why the power supplied is not constant. **(2 marks)**

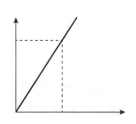

Charging and discharging capacitors

The charge on a charging or discharging capacitor changes exponentially with time.

Discharging a capacitor

Dataloggers can be used to monitor the decay of current and p.d. across a discharging capacitor.

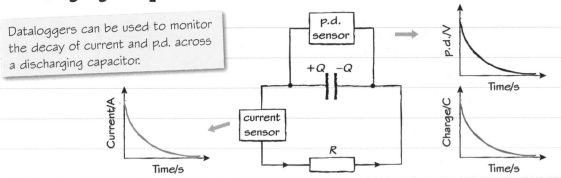

The current, charge and p.d. all decay **exponentially** in the same time. This is because they are all linked by $I = \dfrac{V}{R}$ and $V = \dfrac{Q}{C}$.

Charging a capacitor

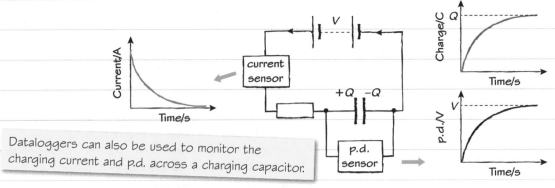

Dataloggers can also be used to monitor the charging current and p.d. across a charging capacitor.

Once again the charge and p.d. curves are the same because $Q = CV$.

The current curve is different – it shows exponential decay. This is because $I = \dfrac{V_R}{R}$, and V_R falls as the p.d. across the capacitor V_C rises. Kirchhoff's second law: $V = V_R + V_C$.

Worked example

Explain why charge initially flows fast from a discharging capacitor but slows down exponentially as capacitor discharges.
(4 marks)

When the capacitor is fully charged, the p.d. across it $V = \dfrac{Q}{C}$ is a maximum, and so the current I (the rate of flow of charge) in the circuit will also be a maximum (if the resistance R in the circuit is constant), $I = \dfrac{V}{R}$. As the charge Q on the capacitor drops, the p.d. across it also drops and hence the current in the circuit drops – in other words, the charge flows more slowly from the capacitor. Numerically, the decrease is exponential: if Q falls by half after a time interval t, the rate of flow of charge is also halved after t. It therefore takes the same interval t for the charge to halve again to $\dfrac{Q}{4}$, and so on.

Now try this

A 50 µF capacitor is fully charged by a 6.0 V supply and then removed and connected into a circuit of resistance 4 kΩ. The current drops to half its initial value in 139 s. Calculate the initial current and the time it will take to drop to one-quarter and one-eighth of that value.
(3 marks)

The time constant

The time taken for a capacitor to charge or discharge can be calculated with the time constant, which depends on the capacitance and the circuit resistance.

Discharging a capacitor

Charge/C vs Time/s

To make charge from a capacitor flow in the circuit for longer:

1 store more charge on the capacitor

2 decrease the rate at which the capacitor discharges.

For the same maximum p.d., increasing the capacitance, C, will increase the charge stored, as $Q = CV$.

Alternatively, increasing the resistance, R, in the circuit, will slow the discharge by decreasing the current, because $\frac{V}{R} = I$. In other words, the time it takes to discharge a capacitor depends on both C and R.

 Practical skills **Using dataloggers**

The curve shown here and those on page 72 can be recorded using dataloggers.

Dataloggers are used with sensors to frequently sample and record data, for example voltage and current in a circuit, into a spreadsheet. Software then displays and analyses the data. In these experiments the voltage sensor might measure the voltage across the capacitor 100 times per second. If this sampling rate is too low it might miss rapid changes and if it is too high the number of data points will be huge, so it can be altered. An alternative method in these experiments uses a storage oscilloscope. This measures the voltage and time and has a memory so that data can be stored and displayed on screen as a graph of voltage against time.

Time constant

The time taken to charge or discharge depends on the **time constant** for the circuit:

$$\tau = RC$$

This is a value in seconds.

The larger the time constant, the longer charging or discharging takes.

During discharge the charge on the capacitor falls to 0.63 ($1/e$) of its original value in a time equal to one time constant.

> Make sure you change to S.I. units before calculating the time constant.

Worked example

(a) Explain why the time taken to charge a capacitor from a supply of p.d. V increases if the circuit resistance is increased. **(2 marks)**

The final charge stored is $Q = CV$, independent of the current or resistance. If resistance is increased, the charging current is smaller so more time is needed to supply this charge. The time constant RC is greater.

(b) Calculate the time constant for a circuit containing a 50 μF capacitor connected in series with a 2.5 kΩ resistor. **(2 marks)**

$\tau = RC = 50 \times 10^{-6} \times 2.5 \times 10^3 = 0.125\,s$

Now try this

1 A 220 μF capacitor is being charged from a 12 V supply. It is fully charged after 5τs?
   (a) Calculate the maximum value for the circuit resistance if the charging process is to be complete in no more than 1.0 s. **(3 marks)**
   (b) Explain why this is a maximum value. **(2 marks)**

2 Show that RC has units of seconds. **(3 marks)**

Exponential decay of charge

The decay of charge on a capacitor can be modelled mathematically as an exponential function.

Understanding the decay equation

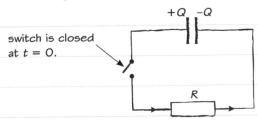

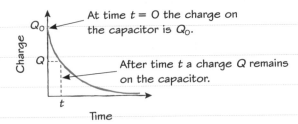

At time $t = 0$ the charge on the capacitor is Q_0.

After time t a charge Q remains on the capacitor.

The decay of the charge Q shown in the graph is represented by this equation:

$Q = Q_0\, e^{-t/RC}$ ———— fraction of original charge left after time t

charge at time t

initial charge at $t = 0$

Current and p.d. during discharge

The p.d. across a capacitor and the discharge current are both directly proportional to its charge, as described on page 71.

Therefore p.d. and discharge current also decay exponentially.

Q, I and V decay with the same time constant:

$Q = Q_0\, e^{-t/RC}$

$V = V_0\, e^{-t/RC}$

$I = I_0\, e^{-t/RC}$

⊞ Maths skills Logarithmic equations

Natural logarithms (written ln) are logs to base e. Just as $\log_{10} 10 = 1$ and $\log_{10} 100 = 2$, $\ln e = 1$ and $\ln e^x = x$.

Taking natural logs of $Q = Q_0\, e^{-t/RC}$:

$\ln Q = \ln(Q_0 \times e^{-t/RC})$

$\ln Q = \ln Q_0 + \ln e^{-t/RC}$

$\ln Q = \ln Q_0 - \dfrac{t}{RC}$

Similarly, $\ln V = \ln V_0 - \dfrac{t}{RC}$

$\ln I = \ln I_0 - \dfrac{t}{RC}$

Worked example

A 2200 μF capacitor is connected to a 6.0 V supply and fully charged. It is then disconnected from the supply and connected across a 10 kΩ resistor.

(a) Calculate the initial discharge current.   **(2 marks)**

$I = \dfrac{V}{R} = \dfrac{6.0}{10\,000} = 6.0 \times 10^{-4}\,\text{A} = 0.60\,\text{mA}$

(b) Calculate the charge on the capacitor after one time constant.   **(3 marks)**

Initial charge $Q_0 = CV_0 = 2.2 \times 10^{-3} \times 6.0$

$= 0.0132\,\text{C}$

Charge remaining after one time constant :

$Q = Q_0\, e^{-t/RC} = Q_0\, e^{-1} = 0.37\, Q_0$

$= 4.9 \times 10^{-3}\,\text{C}$

(c) Calculate the p.d. across the capacitor after 20 s.   **(2 marks)**

$V = V_0 e^{-t/RC} = 6.0\, e^{-(20/(10\,000 \times 0.0022))} = 2.4\,\text{V}$

(d) How much of the stored energy has been dissipated by the time the p.d. across the capacitor has fallen to 3.0 V?   **(2 marks)**

Energy is proportional to p.d. squared:

$W = \frac{1}{2} CV^2$, so if p.d. halves, the remaining stored energy falls to $\frac{1}{4}$ of its original value and therefore $\frac{3}{4}$ (75%) has been dissipated as heat in the resistor.

Now try this

1 Calculate the time constant for a 47 μF capacitor through a 500 Ω resistor and the fraction of charge left on the capacitor after one, two, three, four and five time constants. Comment on how the discharge times would vary if the resistor were replaced with a 200 Ω resistor.   **(6 marks)**

2 A 500 nF capacitor is charged to 3.50 V and then discharged through a 220 Ω resistor.
   (a) Calculate the charge stored on the capacitor.   **(2 marks)**
   (b) Calculate the time constant for the discharge circuit.   **(1 mark)**
   (c) Calculate the charge remaining on the capacitor after 50 μs.   **(4 marks)**

3 Show that when a capacitor of capacitance C discharges through a resistor of resistance R the charge on the capacitor halves during a time equal to $RC \ln 2$.   **(2 marks)**

Exam skills 8

This exam-style question uses knowledge and skills you have already revised. Have a look at pages 70–74, for a reminder about capacitors.

Worked example

A defibrillator is an electrical device sometimes used to provide an electric shock to restart the heart when a patient's heart has stopped beating. The key component is a capacitor. The simplified circuit diagram below shows how a defibrillator can be charged from a power supply and then discharged through the patient.

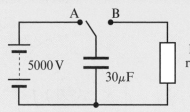

> You might not be familiar with defibrillators, but an unfamiliar context should not put you off. The question is testing your knowledge and understanding of capacitor charging and discharging circuits.

(a) Calculate the energy stored on the capacitor when it is fully charged. **(2 marks)**

$$W = \tfrac{1}{2}CV^2 = \tfrac{1}{2} \times 30 \times 10^{-6} \times 5000^2$$
$$= 375\,J$$

> The energy stored in a charged capacitor can be given by any of three expressions:
> $$W = \tfrac{1}{2}CV^2 = \tfrac{1}{2}QV = \tfrac{1}{2}\frac{Q^2}{C}$$
> It makes sense to use the equation which contains the information provided in the question:
> $W = \tfrac{1}{2}CV^2$.

(b) Calculate the initial discharge current through the patient when the switch is moved from A to B. **(1 mark)**

$$I = \frac{V}{R} = \frac{5000}{120} = 42\,A$$

(c) Calculate the time constant for the discharge circuit and explain its significance. **(3 marks)**

Time constant $\tau = RC = 120 \times 30 \times 10^{-6}$
$$= 0.0036\,s = 3.6\,ms$$

The time constant is the time taken for the charge on the capacitor (and therefore the discharge current) to fall to $\frac{1}{e}$ of its original value (0.37 of its initial value).

(d) Sketch a graph to show how the discharge current varies with time from the moment the switch is moved from A to B.
Show the values of current and charge after two time constants. **(4 marks)**

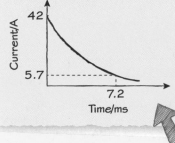

Note how the current at two time constants has been indicated. $2\tau = 7.2\,ms$ and the current is now $42 \times 0.37^2 = 5.7\,A$

Command word: 'Sketch'

'Sketch' does not mean draw a vague diagram or artist's impression of a graph! You should still take care to:

- ✓ Draw the axes with a ruler.
- ✓ Label both axes with quantity and unit.
- ✓ Draw a smooth line or curve.
- ✓ Include values on the axes if you know them.

> What you are not doing is plotting the graph. However, if there are known points that must lie on the graph, these can help you to draw it more accurately. In this case we know the time constant, so we know that the current falls by a factor of 0.37 (e^{-1}) every 3.6 ms.

Describing magnetic fields

Permanent magnets and electric currents create magnetic fields in space.

Magnetic poles

- Like poles repel.
- Opposite poles attract.

A bar magnet has two magnetic poles – north and south. They are examples of magnetic dipoles. Isolated north or south poles, monopoles, do not exist. Each magnet creates a magnetic field that affects the other magnet.

Magnetic flux density, B

The strength of a magnetic field at a point in space, represented by the density of field lines at that point, is called **magnetic flux density**, symbol B, S.I. unit tesla (T).

The Earth's magnetic field near the surface is about 30–60 μT.

A strong permanent magnet might have a flux density of 0.1 T near its pole.

Magnetic flux, F

Magnetic flux indicates the amount of magnetic field passing perpendicularly through a defined area. It has symbol Φ, S.I. unit weber (Wb). $1\,Wb = 1\,T\,m^2$ and $1\,T = 1\,Wb\,m^{-2}$.

Magnetic flux is represented in diagrams by the number of lines passing perpendicularly through a closed loop or area of a magnetic field.

Magnetic flux linkage

If flux Φ passes through a coil, each turn creates a loop with a flux of BA. The flux inside a coil of N turns is therefore NBA. This is the **flux linkage**, $N\Phi$, S.I. units Wb or Wb turns.

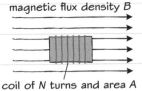

magnetic flux density B

For a coil at 90° to the field.

coil of N turns and area A

- flux through coil $\Phi = BA$
- flux linkage $= N\Phi = NBA$

Rules for drawing magnetic field lines

1. Lines start on north poles and end on south poles, or form closed loops.

2. Lines cannot cross.

3. Lines represent the direction of force experienced by another north pole.

4. Lines are denser where field is stronger.

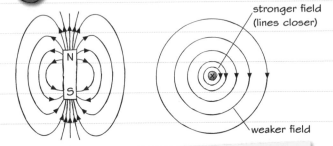

stronger field (lines closer)

weaker field

Magnetic fields of a bar magnet and a straight wire carrying current into the page.

Calculating magnetic flux

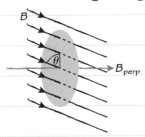

Magnetic flux Φ must be measured perpendicular to area A.

If a uniform magnetic field of flux density B is at angle θ to the area A considered, then the flux perpendicular to A is:

$$\Phi = BA\sin\theta$$

Since $\sin\theta = 1$ when $\theta = 90°$, then if the field is at 90° to the area A the flux is given by:

$$\Phi = BA$$

Now try this

50°

The Earth's magnetic field at a particular place has a flux density of 40 μT and its direction is at 50° to the horizontal.

(a) Calculate the magnetic flux through 1.0 m² of the Earth's surface. **(3 marks)**

(b) (i) A student holds a coil of wire in the magnetic field. How should she hold it in order to get maximum flux linkage? **(1 mark)**

   (ii) Calculate the maximum flux linkage that could be achieved using a square coil of side 0.30 m with 50 turns of wire. **(2 marks)**

   (iii) She now rotates the coil to the position in which the flux linkage is zero. The rotation takes 2.0 s. What is the average rate of change of flux linkage through the coil? **(3 marks)**

Forces on moving charges in a magnetic field

A charge moving in a magnetic field experiences a force.

The magnetic force on a moving charge

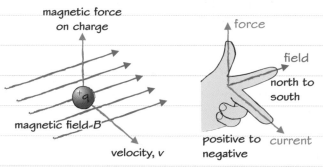

The magnetic force on a charged particle moving in a magnetic field is perpendicular to its velocity and the magnetic flux density. The direction of the force can be predicted using **Fleming's left-hand rule.**

To calculate the magnitude of the magnetic force, F,

$$F = Bqv\sin\theta$$

Changing the sign of the charge changes the direction of the force.

θ is the angle between the velocity of the charged particle and the magnetic field. The magnitude of the force is greatest when θ is 90°, as in the left-hand rule. If the charged particle is moving parallel to the field lines then theta is zero and no magnetic force acts on it.

Magnetic force on a current-carrying conductor

$$F = BIl\sin\theta$$

θ is the angle between the current and the magnetic field

l is the length of current-carrying conductor in the field. Fleming's left-hand rule can be used to predict the direction of magnetic force on a moving charge or an electric current.

Remember: conventional current has the same direction as the movement of positive charges.

Motion of a charged particle in a magnetic field

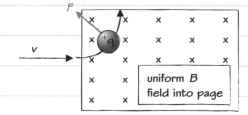

uniform B field into page

The magnetic force is **always** perpendicular to the velocity, so:

- it acts as a centripetal force and the charged particle moves in an arc of a circle
- it does no work on the charged particle so the speed is constant.

Now try this

1 An electric power line carries a current of 5.0 A over a distance of 10.0 km. The mean value of the Earth's magnetic flux density is 35 μT at 70° to the power line. Calculate the magnetic force on the power line. **(3 marks)**

2 Explain why magnetic fields cannot do work on moving charges. **(3 marks)**

3 Calculate the magnetic force on an electron travelling at 3.5×10^6 m s^{-1} parallel to a magnetic field of strength 1.2 T. **(1 mark)**

4 Copy the diagram above that shows the path of a charge $+q$ in a magnetic field. Add lines to show the paths followed by a charge $-q$ injected into the same field:
(a) If the new charge has the same mass and initial velocity as the charge $+q$ **(1 mark)**
(b) If the new charge has the same mass but a higher velocity than the charge $+q$ **(1 mark)**
(c) If the new charge has greater mass but is injected with the same velocity as the charge $+q$. **(1 mark)**

Electromagnetic induction – relative motion

Moving a conductor through a magnetic field, or vice versa, can generate electricity.

Relative motion between a coil and a magnet

A magnet is moved in and out relative to a coil connected to a voltmeter.

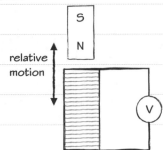

Observations:

1 If there is no motion there is no induced e.m.f.

2 When the coil and magnet move in relation to each other an e.m.f. is induced.

3 If the motion direction changes the sign of the e.m.f. also changes.

4 The faster the motion the greater the induced e.m.f.

5 The effect is the same whether the magnet or the coil is moved.

When there is relative motion between the coil and the magnet, the flux linkage changes. This induces an e.m.f. in the coil. This process is called **electromagnetic induction**.

The size of the induced e.m.f. is directly proportional to the rate of change of flux linkage. This is a version of **Faraday's law**.

Faraday's law is discussed on page 80.

Generator: spinning coil

As the coil rotates, the flux through the coil changes continuously inducing an alternating e.m.f. at the same frequency as the rotation.

Increasing the frequency of rotation increases the amplitude and frequency of the induced alternating e.m.f.

If slip rings are used to connect the rotating coil to a circuit, a.c. is generated. A split-ring commutator will deliver pulsed d.c.

Induced currents

Electromagnetic induction results in an induced e.m.f. Charge can only flow if there is a complete circuit.

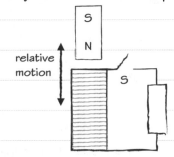

	e.m.f.	Current
Switch open	✓	✗
Switch closed	✓	✓

Generator: rotating magnet

The induced e.m.f. alternates with the same frequency as the rotating magnet.

Increasing the rate of rotation increases the frequency and amplitude of the alternating e.m.f. and thus, in a circuit, the alternating current (a.c.).

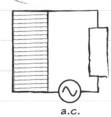

Now try this

A magnet is dropped through a coil and the induced e.m.f. is recorded on a datalogger. The variation of induced voltage with time is shown in the graph.

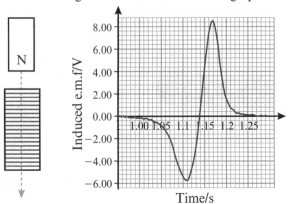

(a) Explain why there are two peaks.   **(2 marks)**

(b) Explain why the peaks have opposite signs.   **(2 marks)**

(c) Explain why the first peak is lower and broader than the second one.   **(4 marks)**

Changing flux linkage

You do not have to move anything to induce an e.m.f.; you can do it by changing the magnetic field.

Changing flux linkage without movement

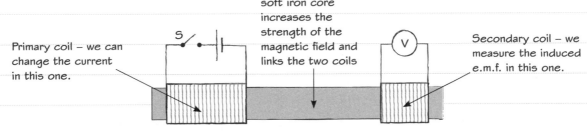

Primary coil – we can change the current in this one.

soft iron core increases the strength of the magnetic field and links the two coils

Secondary coil – we measure the induced e.m.f. in this one.

As S is closed there is a momentary positive pulse of e.m.f. induced in the secondary coil. While S remains closed there is no induced e.m.f. in the secondary coil. As S is opened there is a momentary negative pulse of e.m.f. induced in the secondary coil.

The reason for these observations is that when S is closed the primary coil becomes an electromagnet. Flux from the primary coil passes through the secondary coil.

When the flux changes (closing or opening S) an e.m.f. is induced in the secondary coil.

When the flux is zero or constant there is no induced e.m.f.

Induced e.m.f. is directly proportional to the rate of change of flux linkage through the secondary coil.

The transformer

If an alternating current is applied to the primary coil, the magnetic flux in the core will change continuously. This induces an alternating e.m.f. in the secondary coil. Transformers cannot work with a d.c. input because there is no change of magnetic flux linkage (apart from a momentary one when the device is switched on or off).

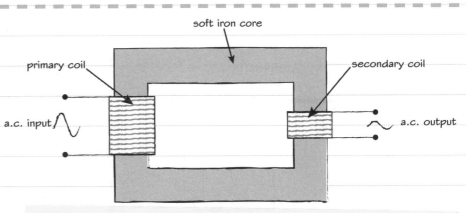

soft iron core

primary coil

secondary coil

a.c. input

a.c. output

Varying the number of turns on the coils can be used to step the original p.d. up or down. This is a step-down transformer, with lower output p.d., as there are fewer turns on the secondary coil.

Changing flux linkage and induced e.m.f.

Time

Flux linkage

Time

Induced e.m.f.

The induced e.m.f. in a coil and the flux linkage through it.

- The maximum induced e.m.f. occurs at the maximum rate of change of flux linkage, steepest gradient.
- The positive peak of the e.m.f. corresponds to the steepest negative slope.

Now try this

1 The ignition coil of a car is connected to a 12 V battery but is capable of producing several kilovolts when the current passing through it is suddenly interrupted. Explain how this is possible. **(3 marks)**

2 (a) If a solid soft iron core is used in a transformer it gets hot and wastes a lot of energy. Explain this effect. **(3 marks)**

(b) In practice the cores of transformers are made from thin sheets of iron that are sandwiched together with electrically insulating layers, a 'laminated core'. Suggest how this might reduce the heat losses in the core. **(3 marks)**

Faraday's and Lenz's laws

All of the effects of electromagnetic induction can be explained using two laws, which can be expressed by a single equation.

The combined equation

Induced e.m.f. (in V) = negative rate of change of flux linkage (Wb s⁻¹)

$$\varepsilon = \frac{-d(N\Phi)}{dt}$$

This combines two laws:

1 **Faraday's law:** the magnitude of the induced e.m.f., ε, is equal to the rate of change of flux linkage, $\frac{d(N\Phi)}{dt}$.

2 **Lenz's law** the direction of an induced e.m.f. is such as to oppose the change creating it. Lenz's law is indicated by the negative sign in the equation and is really a statement of the **law of conservation of energy**. It means that any induced currents will oppose the change that caused them.

Lenz's law

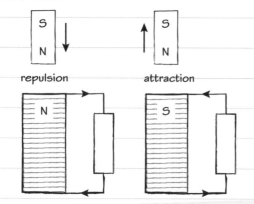

repulsion   attraction

When the magnet is pushed toward the coil the induced currents repel it. When it is pulled away they attract it. In both cases they **oppose** the change causing the induction. This means work must be done to move the magnet. This work transfers energy to the induced current and conserves energy overall. If there was a break in the circuit no current would flow and no energy transfer would take place.

Rate of change of flux linkage

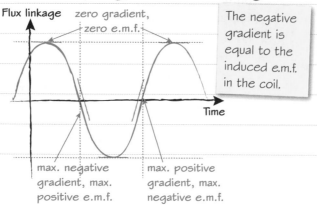

Flux linkage   zero gradient, zero e.m.f.

The negative gradient is equal to the induced e.m.f. in the coil.

Time

max. negative gradient, max. positive e.m.f.

max. positive gradient, max. negative e.m.f.

Rate of change of flux linkage

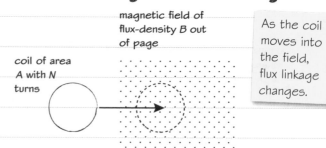

magnetic field of flux-density B out of page

As the coil moves into the field, flux linkage changes.

coil of area A with N turns

Change in flux through coil $\Delta\Phi = +BA$

Change in flux linkage $\Delta(N\Phi) = +NBA$

If change occurs in time Δt, then induced e.m.f.

$$\varepsilon = -\frac{\Delta(N\Phi)}{\Delta t} = -\frac{NBA}{\Delta t}$$

Lenz's law: The flux has increased through the coil out of the page so the direction of current flow is such as to create a field into the page, opposing the change.

Maths skills $\frac{\Delta t}{\frac{\Delta(N\Phi)}{\Delta t}}$ means the change in $N\Phi$ in a small time Δt, so it gives the average rate of change in this time.

As Δt becomes smaller, the average rate of change becomes the instantaneous rate of change. As Δt approaches zero we call it dt, which represents an infinitesimal change in time. $\frac{d(N\Phi)}{dt}$ thus means the instantaneous rate of change of flux linkage.

Now try this

A coil of 40 turns and cross-sectional area 0.016 m² is placed with its plane perpendicular to a uniform magnetic field of strength 0.075 T. It is slowly rotated so that its plane is now parallel to the same field. The time of rotation is 5.0 s and the resistance of the coil is 8.0 Ω.

(a) Calculate the initial flux linkage and the flux linkage when the plane of the coil has rotated through 90°. **(3 marks)**

(b) Calculate the average rate of change of flux linkage in the coil. **(2 marks)**

(c) State the average induced e.m.f. in the coil during the rotation. **(1 mark)**

(d) Calculate the average induced current in the coil. **(2 marks)**

Alternating currents

Sources of e.m.f. such as cells and batteries have constant current direction; the mains supply, however, alternates.

Direct and alternating current, DC and AC

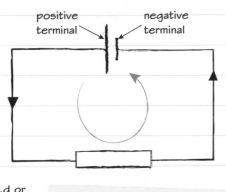

positive terminal / negative terminal

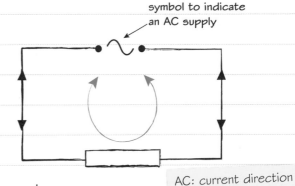

symbol to indicate an AC supply

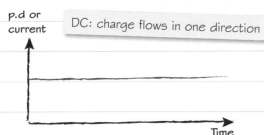

p.d or current

DC: charge flows in one direction

Time

p.d or current

AC: current direction changes periodically

Time

Describing an AC p.d.

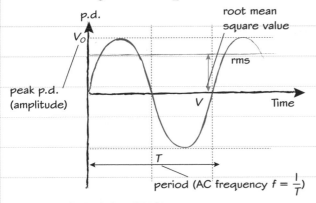

p.d.
V_0
root mean square value
rms
peak p.d. (amplitude)
V
Time
T
period (AC frequency $f = \frac{1}{T}$)

Root mean square (rms) values

Maths skills

To calculate the root mean (rms) value of AC or p.d. simply divide the peak values by $\sqrt{2}$:

$$V_{rms} = \frac{V_0}{\sqrt{2}} \text{ and } I_{rms} = \frac{I_0}{\sqrt{2}}$$

The rms value of an AC p.d. is its DC **equivalent**. This means that 12 V rms AC will provide the same power as a steady DC at a p.d. of 12 V.

RMS values of p.d. and current can be used in the same equations for energy and power as are used for DC calculations.

Worked example

The UK mains supply has an rms p.d. of about 230 V and a frequency of 50 Hz.

(a) Calculate the peak value of the UK mains p.d. **(1 mark)**

$V_0 = \sqrt{2} \times 230 = 325\,V$.

(b) Calculate the period of the UK mains supply. **(1 mark)**

$T = \frac{1}{f} = \frac{1}{50} = 0.020\,s = 20\,ms$

(c) Calculate the peak current in a 100 W lamp connected to the UK mains supply. **(2 marks)**

$P = I_{rms}V_{rms}$ so $I_{rms} = \frac{P}{V_{rms}} = \frac{100}{230} = 0.43\,A$.

Therefore peak current $= \sqrt{2} \times 0.43$
$= 0.61\,A$

Now try this

An AC power supply of peak value 16.0 V and period 50 ms is connected to a resistor of constant resistance 50 Ω.
(a) Calculate the rms p.d. of the supply. **(1 mark)**
(b) Calculate the frequency of the supply. **(1 mark)**
(c) Calculate the rms current supplied to the resistor. **(2 marks)**
(d) Calculate the power dissipated in the resistor. **(2 marks)**

Exam skills 9

This exam-style question uses knowledge and skills you have already revised. Have a look at pages 76–81, for a reminder about magnetic flux linkage and Faraday's and Lenz's laws.

Worked example

The diagram shows a simple experiment used to demonstrate electromagnetic induction.

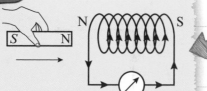

As the bar magnet is moved toward one end of the coil the ammeter indicates a small current flowing around the circuit.

(a) Explain how the current is generated. **(3 marks)**

- The bar magnet creates a magnetic field in the space around it.
- As the bar magnet moves toward the coil, the magnetic flux through the coil changes.
- Faraday's law states that there will be an induced e.m.f. in the coil equal to the rate of change of magnetic flux linkage.
- The circuit is complete, so the e.m.f. results in a flow of current.

(b) State and explain the effect of reversing the direction of motion of the bar magnet and moving it away from the coil at a higher speed. **(3 marks)**

The direction of current in the circuit would reverse and the magnitude of the current would increase.

The current reverses because the flux linkage is now reducing, not increasing. This changes the sign of the rate of change of flux linkage and so reverses the sign of the induced e.m.f.

The current increases in magnitude because the rate of change of flux linkage is greater.

(c) When the bar magnet is moved toward the coil, the flux through the circuit changes by 8.0×10^{-4} Wb in 0.50 s. The coil has 20 turns and the total resistance of the circuit is 5.0 W. Calculate the average induced current during this time. **(3 marks)**

Change of flux linkage = $8.0 \times 10^{-4} \times 20$
$= 0.016$ Wb

Rate of change of flux linkage = $\dfrac{0.016}{0.50}$

$= 0.032$ Wb s^{-1}
$=$ induced e.m.f.

Average current $= \dfrac{\text{induced e.m.f.}}{\text{resistance}} = \dfrac{0.032}{5.0}$

$= 6.4 \times 10^{-3}$ A
$= 6.4$ mA.

The introduction to this question mentions electromagnetic induction – this should alert you to the need to use Faraday's and Lenz's laws and to think about magnetic field strength and flux.

Notice how the answer has been laid out in bullet points – this is often a useful strategy when answering a question that is worth several marks. It helps to ensure that you are making separate points. It also helps to set up a logical structure to your answer. The answer to part (c) has been presented in the same way.

When giving explanations try to include relevant technical terms – this helps to show that you understand the important underlying physics.

Read the question carefully and make sure you answer all parts of it. In this question there are two things to state and two explanations to give.

An alternative approach to answering (b) would be to start by quoting Faraday's and Lenz's laws in the form:

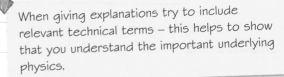

$$E = -\frac{d(N\Phi)}{dt}$$

and then indicating how the changes affect this:

- reversing direction changes the sign of the change
- increasing speed increases the rate of change
- both affect the induced e.m.f. directly.

Notice how this has been structured – whilst it would be acceptable to quote Faraday's law and substitute all the values straight into the equation, it is often helpful to break up the calculation into simple steps.

The Rutherford scattering experiment

This experiment changed our model of the atom and suggested that solid objects are almost entirely empty space!

The alpha-particle scattering experiment

Rutherford used a radioactive source to fire alpha particles at very thin gold foil. To his surprise, although most of them went straight through with very small deflections, a small number were deflected through large angles and some even bounced back off the foil.

Alpha scattering

As alpha particles approach the gold nucleus a repulsive electrostatic force deflects them.

Most alpha particles are barely deflected, but a few are deflected right back.

Alpha particles undergoing a head-on collision get closest to the nucleus.

Rutherford used the closest approach to calculate an upper limit for the size of the nucleus. Nuclear radii are more than 10^4 times smaller than atomic radii.

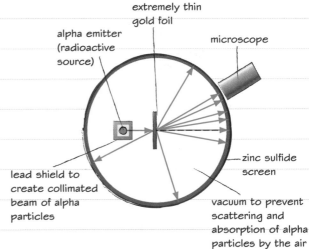

extremely thin gold foil
alpha emitter (radioactive source)
microscope
lead shield to create collimated beam of alpha particles
zinc sulfide screen
vacuum to prevent scattering and absorption of alpha particles by the air

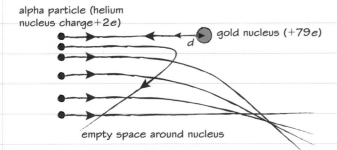

alpha particle (helium nucleus charge +2e)
gold nucleus (+79e)
d
empty space around nucleus

The Rutherford model of the atom

Before the experiment atoms were considered to be small hard spheres of positive matter with negatively charged electrons stuck into them – the 'plum-pudding' model.

Rutherford concluded from his results that:

1 The atom is mainly empty space.

2 There is a tiny central nucleus with distant orbiting electrons.

3 The nucleus must contain nearly all of the mass of the atom and be charged.

Rutherford's 'planetary model' of the atom

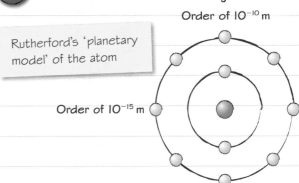

Order of 10^{-10} m

Order of 10^{-15} m

Worked example

Explain how the Rutherford model accounts for the following observations in the alpha scattering experiment:
(a) Most alpha particles undergo very small amounts of scattering. **(2 marks)**

The nucleus is tiny compared with the atom so most alpha particles do not pass close to a nucleus. This means that the electrostatic repulsion between the nuclei and most alpha particles is tiny so they do not deflect much.

(b) A small fraction of the incident alpha particles scatter through more than 90°. **(3 marks)**

A small proportion of the incident alpha particles will pass very close to a nucleus. They experience a large repulsive electrostatic force because both the alpha particle and the nucleus are positively charged. The nucleus is much more massive than the alpha particle so it barely moves but the alpha particle is deflected through a large angle.

Now try this

1 Explain why Rutherford concluded that:
   (a) The atom is mainly empty space. **(2 marks)**
   (b) Most of the mass of the atom is in its nucleus. **(2 marks)**
   (c) The nucleus must be charged **(3 marks)**

2 Solid matter has a typical density of about 5000 kg m^{-3}; suggest a value for the density of nuclear matter. Explain how you obtained your value. **(3 marks)**

Nuclear notation

The nuclear model of the atom helps us to understand the essential differences between atoms of different elements.

The structure of the nucleus

particles in nucleus: } proton: charge $+e$　nucleus held together by strong nuclear
nucleons } neutron: neutral　force – overcomes repulsion between protons

nucleon (mass) number A
(= number of protons + neutrons)

$^{7}_{3}\text{Li}$ lithium-7 nucleus has 3 protons and 4 neutrons (the atom has 3 electrons)

proton number Z
(determines the element)

in a neutral atom, number of electrons = number of protons = Z
number of neutrons in the nucleus $N = A - Z$

Nuclear transformations

Nuclei can change in three ways:

- radioactive decay
- nuclear fission
- nuclear fusion.

You can review these types of nuclear transformation on pages 110, 111 and 113.

Nuclear transformations require or release far more energy per atom than chemical reactions because the strong nuclear forces between nucleons are far greater than the forces that bind the electrons to the nucleus and form chemical bonds.

In a nuclear transformation:

- total number of protons + neutrons is unchanged
- total charge is unchanged.

These **conservation laws** are used to balance nuclear transformation equations. For example, in the alpha decay – the emission of a helium nucleus from an unstable heavy nucleus – of uranium-238:

$$^{238}_{92}\text{U} \rightarrow \,^{234}_{90}\text{Th} + \,^{4}_{2}\text{He}$$

The top line of nucleon numbers balances:

$238 = 234 + 4$

The bottom line of charge balances:

$92 = 90 + 2$

Isotopes

Isotopes are atoms with the same atomic number (Z) but different nucleon numbers (A). They are the same element with the same chemical properties but a different number of neutrons in the nucleus.

Some isotopes of carbon:

$^{12}_{6}\text{C}$　$^{13}_{6}\text{C}$　$^{14}_{6}\text{C}$

Z	6	6	6 protons
A	12	13	14 nucleons
N	6	7	8 neutrons

Carbon-12 and carbon-13 are stable isotopes. Carbon-14 is a radioactive beta-minus emitter used in radiocarbon dating.

Some isotopes of uranium:

$^{235}_{92}\text{U}$　$^{238}_{92}\text{U}$

Z	92	92 protons
A	235	238 nucleons
N	143	146 neutrons

Uranium-238 makes up 99.3% of all natural uranium. Uranium-235, the remainder, is fissile.

Now try this

1 The alpha decay of uranium-238 shown above results in the formation of a thorium-234 nucleus. State the number of:
(a) nucleons **(1 mark)**
(b) protons **(1 mark)**
(c) neutrons in the thorium-234 nucleus. **(1 mark)**
(d) State the number of electrons in a neutral thorium-234 atom. **(1 mark)**

2 It is almost impossible to separate uranium-235 from uranium-238 by chemical reactions. Suggest a reason for this. **(3 marks)**

3 Write down the symbols for the isotopes nitrogen-14, nitrogen-15 (both with 7 protons), potassium-39 and potassium-41 (19 protons). **(3 marks)**

Worked example

Oxygen-15 is a radioactive isotope of oxygen. Its nucleus is represented by:

$^{15}_{8}\text{O}$

State how many nucleons, protons and neutrons there are in an oxygen-15 nucleus, and how many electrons orbit the nucleus in the neutral oxygen-15 atom. **(4 marks)**

$A = 15, Z = 8, N = Z - A = 7$; 8 electrons.

Electron guns and linear accelerators

Electric and magnetic fields are used to accelerate, focus and direct charged particles in particle accelerators.

Thermionic emission

When a metal is heated the electrons inside it gain energy. This enables some of them to leave the surface of the metal. An electric field can then accelerate them into a beam.

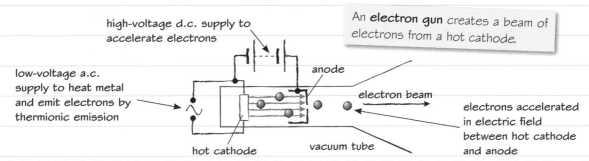

high-voltage d.c. supply to accelerate electrons

An **electron gun** creates a beam of electrons from a hot cathode.

anode

low-voltage a.c. supply to heat metal and emit electrons by thermionic emission

electron beam

electrons accelerated in electric field between hot cathode and anode

hot cathode

vacuum tube

Velocity of electrons

The energy transferred to the electrons as they are accelerated through the p.d. is equivalent to the charge × p.d., eV. By conservation of energy: $eV = \frac{1}{2}mv^2$ (at non-relativistic speeds), so $v = \sqrt{(2eV/m)}$. For more detail on non-relativistic speeds see page 86.

Linear accelerators (LINACs)

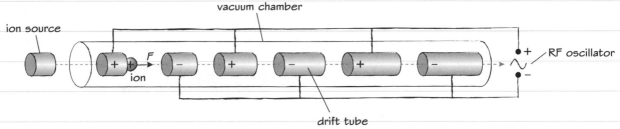

vacuum chamber

ion source

+ RF oscillator

ion

drift tube

Ions can be accelerated by electric fields in a LINAC. An AC supply is used so that the electric field between adjacent cylindrical electrodes (called 'drift tubes') is always in the right direction to accelerate the ions. The drift tubes get longer further along the LINAC, because the AC supply has constant frequency and the ions travel further during each cycle of the AC as they speed up.

Worked example

An electron gun is used to produce an electron beam inside a vacuum tube. The potential difference between the anode and cathode is 500 V. The charge on an electron is 1.60×10^{-19} C and it has a mass of 9.11×10^{-31} kg.

(a) Why is it important to use a vacuum tube? **(2 marks)**

If the tube was filled with air the electrons would scatter from air molecules and change direction. They would also lose energy in the collisions and would ionise the air molecules. No beam would be formed.

(b) Calculate the final velocity of electrons in the beam. **(2 marks)**

$v = \sqrt{(2eV/m)}$

$= \sqrt{\dfrac{2 \times 1.6 \times 10^{-19} \times 500}{9.1 \times 10^{-31}}}$

$= 1.3 \times 10^7\,\text{m s}^{-1}$

Now try this

(a) Explain why the electrons in an electron gun accelerate between the cathode and the anode. **(3 marks)**

(b) Calculate the final velocity of electrons accelerated by a potential difference of 400 V. **(2 marks)**

(c) By what factor does the kinetic energy of the electrons change if the accelerating voltage is doubled? **(2 marks)**

(d) By what factor does the linear momentum of the electrons change if the accelerating voltage is doubled? **(2 marks)**

Cyclotrons

If charged particles are made to move in a circle they can be accelerated on every circuit.

Cyclotrons

Circular electromagnets create a vertical magnetic field to deflect the charged particles into circular arcs.

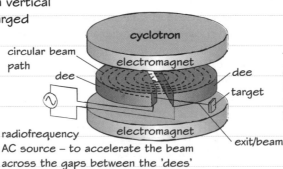

A cyclotron accelerates charged particles using magnetic and electric fields.

radiofrequency AC source – to accelerate the beam across the gaps between the 'dees'

The two 'dees' are connected to an AC supply. Every time the charged particle crosses the gap it is accelerated.

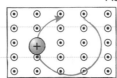

Magnetic forces cause charged particles to move in circular arcs.

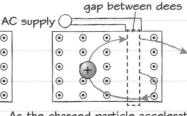

B-field out of page

As the charged particle accelerates the radius of curvature increases.

Charged particles are accelerated in a circle by the magnetic field, and their linear velocity is increased by the alternating electric field in the gap between the 'dees'. The maximum velocity achievable in a cyclotron is limited by the strength of the magnetic field and the diameter of the electromagnets.

The force on charged particles moving in a magnetic field

The radius r of the circular path followed by a charged particle in a uniform magnetic field can be derived from the equation for the force on a charged particle moving across a magnetic field:

$$F = Bqv \sin \theta$$

This force acts perpendicular to the velocity, so $\sin \theta = 1$, and the particle will follow a circular path.

The equation for centripetal force is:

$$F = \frac{mv^2}{r}$$

$$Bqv = \frac{mv^2}{r}$$

$$r = \frac{mv}{Bq}$$

$$r = \frac{P}{Bq}, \text{ often written as } r = \frac{P}{BQ}$$

where m is the mass of the charged particle, v is its velocity, B is the magnetic flux density and q (or Q) is the charge on the particle.

For more on $F = Bqv \sin \theta$ see page 77; for centripetal force $F = mv^2/r$ see page 65.

Worked example

The Diamond Light Source in Oxfordshire is a particle accelerator in which electrons travel around a ring of circumference 560 m. Calculate the magnetic flux density B required to accelerate an electron travelling around the ring at 10^7 m s^{-1}.
($m_e = 9.11 \times 10^{-31} \text{ kg}$, $e = -1.60 \times 10^{-19} \text{ C}$) **(3 marks)**

Circumference of ring $= 2\pi r$ so $r = \dfrac{560}{2\pi}$

$$r = \frac{P}{BQ} = \frac{mv}{BQ}$$

$$B = \frac{mv}{Qr}$$

$$B = \frac{9.11 \times 10^{-31} \times 10^7 \times 2\pi}{1.60 \times 10^{-19} \times 560}$$

$$= 6.4 \times 10^{-7} \text{ T}$$

In fact, electrons travel much faster than this in particle accelerators, and relativistic mass increases have to be taken into account in calculations.

Now try this

1 Charged particles must be accelerated before they are injected into the centre of a cyclotron. Why? **(2 marks)**

2 Why is the maximum energy of a charged particle limited by the diameter of the cyclotron? **(2 marks)**

Particle detectors

If charged particles enter a magnetic field, they will be accelerated into a curving track that depends on their mass and their charge.

Detecting particles by ionisation

Charged particles can transfer energy to the medium, usually a gas, through which they move by ionising it. This gives us a means to detect the charged particles, used in:

- spark chambers
- scintillators
- Geiger–Müller tubes
- ionisation chambers
- cloud and bubble chambers
- calorimeters.

The Geiger–Müller tube

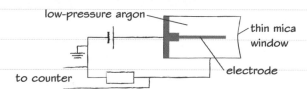

low-pressure argon — thin mica window

to counter — electrode

A Geiger counter detects ionisation events in the low-pressure gas in the tube.

Particle tracks in detectors

Some detectors can show the paths of charged particles, from which it is possible to work out the characteristics of the particle:

- Radius of curvature:
  larger radius = more momentum

On page 86 we saw that $r = mv/Bq$ so the radius of particles with the same charge, q, moving in a uniform magnetic field, B, is proportional to mv, the momentum, of the particles.

- Direction of curvature: sign of charge, from Fleming's left-hand rule.

Look back to page 77 to remind yourself of Fleming's left-hand rule.

Worked example

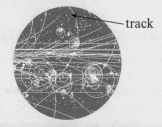

— track

In this hydrogen bubble chamber image, the magnetic field is pointing into the page.
Work out the sign of the charge on the particle that left the marked track. **(2 marks)**

Fleming's left-hand rule, pointing the first finger into the page, shows that the track belongs to a positively charged particle.

Now try this

1 The image shows the paths of two charged particles A and B entering the same magnetic field (directed into the page) at the same velocity. What are the possible conclusions about the mass and charge of the two particles? **(4 marks)**

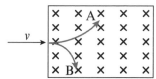

2 The bubble track shown in the worked example above is a spiral. Explain what is happening to the particle causing this track to result in this spiral path. **(3 marks)**

Matter and antimatter

Mass and energy

Einstein's famous equation tells us that energy E and matter are equivalent and can be interconverted. Particles can be created from energy and energy is released if particles are destroyed.

$$\Delta E = c^2 \Delta m$$

c is the speed of light; c^2 is a huge number.

The mass of a particle at rest is called the 'rest mass' m_0. Moving particles have kinetic energy as well as rest energy so their total energy is greater.

Conservation laws

Energy is conserved in all physical interactions. However, energy that is locked up in mass can be released, for example as kinetic energy when a stationary nucleus undergoes alpha decay. The total energy, including the energy in mass, rest mass, is conserved.

Pair creation and annihilation

It is not possible to create a single particle or antiparticle in isolation as this would violate several conservation laws. For example, creating an electron would violate the law of conservation of charge. However, an electron–positron pair can be created if enough energy is available, for example in the form of a photon.

This bubble chamber track shows the creation of two electron-positron (e^-/e^+) pairs by gamma photons at A and B (Gamma photons do not create tracks). In each case the tracks are identical, apart from the direction of curvature, resulting from the opposite charge on each particle. The gamma photon creating the e^-/e^+ pair at B has more energy so the tracks have a greater radius of curvature.

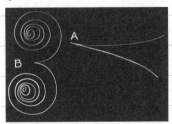

Electron–positron pair production by gamma rays, unseen, in a bubble chamber. A magnetic field is applied, into the page.

The minimum energy needed to create an electron–positron pair is equivalent to the mass of the pair: $\Delta E = 2c^2 m_e$.

If a particle meets its antiparticle, the two annihilate and emit a pair of gamma rays that carry away the energy, each with $E_{photon} = \dfrac{hc}{\lambda} = c^2 m_e$.

The discovery of antimatter

Antimatter was predicted from theory before it was discovered in particle tracks. Every particle has a corresponding antimatter partner or 'twin':

- electron/positron (anti-electron)
- proton/antiproton
- neutron/antineutron
- quark/antiquark, and so on.

Some particles, e.g. photons, are their own antiparticles.

Antimatter particles have:

- the same mass as the corresponding particles
- opposite charge
- opposite quantum numbers.

magnetic field into page

thin lead barrier

Curvature increases (losing momentum as it interacts with barrier) showing that particle is moving up page

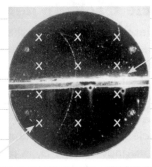

Particle track in a cloud chamber showing the track of a positively charged particle, Fleming's left-hand rule. In every other respect the path is identical to that which would have been made by an electron. It is a positron.

1 Imagine 1.0 kg of antimatter could be safely annihilated with 1.0 kg of matter.
   (a) Calculate the total energy released. **(3 marks)**
   (b) A typical coal-fired power station is 33% efficient and generates about 1.0 GW of electrical power. Calculate for how long the energy from the annihilation could supply the power station. **(4 marks)**

2 (a) Suggest why particle–antiparticle annihilation results in two gamma ray photons rather than one. **(2 marks)**
   (b) Explain what happens when a gamma-ray photon with more than the minimum energy required to create an electron–positron pair does actually produce such a pair. **(2 marks)**

The structure of nucleons

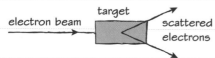

Nucleons can be analysed using particle beams.

Finding the diameter of a nucleon

electron beam → target → scattered electrons

If a high-energy electron beam is directed at a target containing many protons, e.g. liquid hydrogen, the electrons scatter from the protons.

The scattering pattern can be used to analyse the size and structure of the protons. In order to resolve detail, the electron de Broglie wavelength must be comparable to the size of the protons.

$$\lambda = \frac{h}{mv} \approx \text{size of object}$$

If the electrons have enough energy to have a de Broglie wavelength roughly the size of the proton, they will form a diffraction pattern.

Electrons do not experience the strong nuclear force, so they give a better idea of the size of a nucleus than values obtained using alpha particles.

Particles have a wavelength, the de Broglie wavelength, as described on page 53.

Inside the nucleon

If electrons are accelerated to still higher energies, their de Broglie wavelengths become short enough to be used to resolve the internal structure of the nucleon.

LINAC   electron beam   target and detectors

The Stanford Linear Accelerator fired high-energy electrons at a hydrogen target and obtained evidence for the quark structure of nucleons.

The quark structure of nucleons

The proton and neutron are made from **up** and **down** quarks.

 up quark (u): charge $+\frac{2}{3}e$

 down quark (d): charge $-\frac{1}{3}e$

proton: uud,   neutron: udd,
charge $+e$ neutral

Relativistic effects

Particles inside particle accelerators have velocities very close to the velocity of light. At such high velocities relativistic effects are important. Time dilation results in the slowing of time for a moving particle. This means that unstable particles with a short lifetime actually survive for a much longer time in the laboratory if they are moving rapidly. This can help physicists identify them, because they leave longer tracks in particle detectors.

A similar effect is seen for muons produced by cosmic rays hitting the upper atmosphere: far more are recorded at the Earth's surface than might be expected given their half-life of about $1.5\,\mu s$, because they travel at relativistic speeds and thus experience time dilation.

Worked example

Explain why electrons must be accelerated to very high energies if they are to be used to probe the quark structure of a proton. **(3 marks)**

The diameter of a proton is of the order of 10^{-15} m so the electron wavelength must be smaller than this in order to resolve its detailed structure. The wavelength of an electron is given by the de Broglie equation: wavelength = Planck constant / momentum.

To obtain a very short wavelength the electrons must have a very high momentum and therefore a very high energy.

Now try this

Account for the overall charge on the following particles in terms of their quark structure: proton, neutron, pi⁺ meson (up, antidown), pi⁻ meson (down, antiup). **(4 marks)**

Nuclear energy units

It is often convenient to use non-S.I. units when calculating nuclear energy changes, but it is important to be able to convert them to S.I. units!

The electronvolt

An electronvolt is equal to the energy transfer when a charge e is moved through a p.d. of 1.0 V.

1 electronvolt (eV) = 1.60×10^{-19} J

1.0 MeV = 1.0×10^6 eV

1.0 GeV = 1.0×10^9 eV

Masses in MeV/c^2 and GeV/c^2

Mass and energy are related by $\Delta E = c^2 \Delta m$

or $m = \dfrac{E}{c^2}$

Particle physicists often use the units MeV/c^2 and GeV/c^2 as non-S.I. units of mass.

1 MeV/c^2 = 1.8×10^{-30} kg

1 GeV/c^2 = 1.8×10^{-27} kg

In these units, mass of:

electron = 0.511 MeV/c^2

proton = 938 MeV/c^2, almost 1 GeV

neutron = 940 MeV/c^2, almost 1 GeV

The LHC

The Large Hadron Collider (LHC) at CERN is designed to collide particles at energies up to 14 TeV (14 × 10^{12} eV). This means that each collision provides enough energy to create up to a million particles with the rest mass of a proton.

Inside the tunnel at the LHC. The accelerator beam tube is buried underground near the Swiss-French border and forms a ring 27 km in circumference. The beams are accelerated by electric fields and deflected and focused by magnetic fields.

Worked example

1 An electron and a proton are both accelerated through a potential difference of 2000 V.
   (a) Calculate the energy gained in each case and express your answer in both electronvolts and joules. **(3 marks)**

Both gain the same energy because both have a single charge e and are accelerated through the same p.d. Energy gained = 2000 eV (2 keV). Converting to joules:

2000 eV = 2000 × 1.60×10^{-19}
     = 3.2×10^{-16} J

   (b) Calculate the final velocity of each particle. (electron mass = 9.11×10^{-31} kg, proton mass = 1.67×10^{-27} kg) **(2 marks)**

$E_k = \frac{1}{2}mv^2$ so $v = \sqrt{2(E_k)/m}$

For the electron:

$v = \sqrt{\dfrac{2 \times 3.2 \times 10^{-16}}{9.11 \times 10^{-31}}}$

= 2.7×10^7 m s^{-1}

For the proton:

$v = \sqrt{\dfrac{2 \times 3.2 \times 10^{-16}}{1.67 \times 10^{-27}}}$

= 6.2×10^5 m s^{-1}

2 Convert the electron mass of 9.11×10^{-31} kg to MeV/c^2 using $E = mc^2$. **(2 marks)**

First convert the mass to its equivalent energy.

$E = mc^2 = 9.11 \times 10^{-31} \times 9.00 \times 10^{16}$ J
  = 8.20×10^{-14} J = 5.12×10^5 eV

This is the rest energy of the electron in eV.

Now try this

1 Show that the mass of a proton, 1.67×10^{-27} kg, is roughly equivalent to 1 GeV/c^2. **(2 marks)**

2 Through what voltage must an electron be accelerated in order that its kinetic energy is equal to its rest energy? **(3 marks)**

3 Calculate the velocity of a proton accelerated through a potential difference of 50 V. **(3 marks)**

The Standard Model

The wide range of particles and forces in nature can be almost entirely explained using the Standard Model.

The fundamental particles of nature

In the Standard Model, there are three generations of particles of matter:

increasing mass

Generation	Quarks		Leptons	
1	up ($+\frac{2}{3}e$)	down ($-\frac{1}{3}e$)	electron ($-1e$)	electron neutrino (0)
2	charm ($+\frac{2}{3}e$)	strange ($-\frac{1}{3}e$)	muon ($-1e$)	muon neutrino (0)
3	top ($+\frac{2}{3}e$)	bottom ($-\frac{1}{3}e$)	tau ($-1e$)	tau neutrino (0)

Ordinary matter is composed entirely of particles from the first generation: up and down quarks plus electrons. Every quark and lepton has its own antiparticle.

Particles of matter interact by exchanging **force carriers**. The photon is the force carrier responsible for electromagnetic interactions.
- Charged particles interact by electromagnetic forces.
- Particles containing quarks – baryons and mesons – interact by the strong nuclear force.
- Leptons are not affected by the strong nuclear force.

Baryons

Baryons (Greek 'heavy') are made from quark triplets. The charge on a baryon is the sum of the charges on the quarks it contains. For example,
- proton: uud, $+\frac{2}{3} + \frac{2}{3} - \frac{1}{3} = +1$
- neutron: udd, $+\frac{2}{3} - \frac{1}{3} - \frac{1}{3} = 0$

Many other, heavier baryons can be created in particle accelerators, such as the lambda-zero, Λ^0, uds.

Conservation of baryon number

✓ Baryons have a baryon number +1.

✓ Antibaryons have a baryon number −1.

✓ Total baryon number is conserved in all particle interactions.

Leptons

All leptons (Greek 'light') are fundamental particles.

Conservation of lepton number>

✓ Leptons have a lepton number +1.

✓ Antileptons have a lepton number −1.

✓ Total lepton number is conserved in all particle interactions.

Mesons

Mesons (Greek 'middle') are made from quark–antiquark pairs. For example,
- pi-plus (π^+): up + antidown, u$\bar{d}$, $+\frac{2}{3} + \frac{1}{3} = +1$
- pi-minus (π^-): down + antiup, d$\bar{u}$, $-\frac{1}{3} - \frac{2}{3} = -1$

The pi-minus is the antiparticle to the pi-plus. There is also a pi-zero (π^0), which can be made from either the u$\bar{u}$ or the d$\bar{d}$ combination, and is its own antiparticle.

The baryon number of all mesons and antimesons is zero.

Discovery of the top quark

Physicists predicted the existence of the top quark from symmetries in the Standard Model. By predicting its mass they knew how much energy would be needed to create a top quark. It was discovered at the proton–antiproton collider at Fermilab in 1995.

Worked example

Use the quark model to show that the lambda-zero (Λ^0, uds) is neutral. **(2 marks)**

By adding the charges on up, down and strange quarks we get:

$+\frac{2}{3}e - \frac{1}{3}e - \frac{1}{3}e = 0$ (no charge)

Now try this

1 State the charges on each of the following baryons:
   (a) A sigma particle composed of uus quarks. **(2 marks)**
   (b) A double charmed omega particle composed of scc quarks. **(2 marks)**

2 Leptons are regarded as fundamental particles but baryons are not – explain. **(2 marks)**

Particle interactions

When particles interact, the transformations they can undergo are governed by a strict set of conservation laws.

Conservation laws for interactions

 conservation of energy

 conservation of momentum

 conservation of charge

 conservation of baryon number

 conservation of lepton number

All other transformations are forbidden.

Beta decay of a neutron

In radioactive beta decay, a neutron inside a nucleus decays to a proton, an electron and an electron antineutrino:

$$^1_0n \rightarrow {}^1_1p + {}^0_{-1}e + {}^0_0\bar{\nu}$$

 The neutron has greater mass than the proton. When it transforms into a proton, this **mass deficit** is converted to the energy needed to create the electron and neutrino.

2 The sum of momentum vectors for the three particles created in this decay must equal the initial momentum vector of the neutron.

3 The bottom line is like the proton number and tells us the charge: $0 = 1 - 1 + 0$.

4 The top line is the nucleon number or the baryon number: $1 = 1 + 0 + 0$.

5 A lepton and antilepton are created so the lepton number is conserved: $0 = 0 + 1 - 1$. The underlying process here is the change of a down quark into an up quark in the neutron.

Pair creation

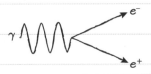

Gamma ray → electron + positron
$$^0_0\gamma \rightarrow {}^0_{-1}e + {}^0_1e$$

The conservation laws mean that:

1 The energy of the gamma ray (hf) must be at least equal to $2m_ec^2$.

2 The sum of vector momenta of the electron and positron is equal to the initial momentum of the gamma-ray photon.

3 The gamma ray photon is uncharged and the sum of charges on the electron and positron is zero: $0 = -1 + 1$.

4 The baryon number is the same before and after the event. It is 0, as no baryons are involved.

5 The lepton number is the same before and after the event: the photon lepton number is zero, and the electron and positron have lepton numbers +1 and −1 so they add to zero.

Worked example

High-energy collisions between protons can create antiprotons. For example:

$$^1_1p + {}^1_1p \rightarrow {}^1_1p + {}^1_1p + {}^1_1p + {}^{-1}_{-1}\bar{p}$$

(a) How can energy be conserved in this reaction? **(2 marks)**

The incoming protons are accelerated to high energies, so some of their kinetic energy creates the rest energy of the two new particles.

(b) How is baryon number conserved in this reaction? **(2 marks)**

Left: baryon number = 1 + 1 = 2

Right: baryon number = 1 + 1 + 1 − 1 = 2

(c) How is charge conserved in this reaction? **(2 marks)**

Left: charge = 1 + 1 = +2

Right: charge = 1 + 1 + 1 − 1 = +2

Now try this

1 A suggested reaction to create antiprotons is:

$$^1_1p + {}^1_1p \rightarrow {}^1_1p + {}^1_1p + {}^0_1\pi^+ + {}^{-1}_{-1}\bar{p}$$

  (a) Show whether this reaction conserves charge. **(2 marks)**

  (b) Show whether this reaction conserves baryon number. **(2 marks)**

  (c) State whether the reaction is permitted. **(1 mark)**

2 Explain how lepton number is conserved during electron–positron annihilation to form two gamma ray photons. **(3 marks)**

Exam skills 10

This exam-style question uses knowledge and skills you have already revised. Have a look at pages 77, 85 and 86, for a reminder about charged particles in magnetic fields, electron guns and particle accelerators.

Worked example

The electron deflection tube shown consists of an electron gun inside a vacuum tube. Two parallel metal plates inside the tube can be used to create a variable electric field, and two external coils can be used to create a variable magnetic field.

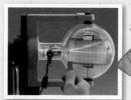

A fluorescent screen shows up the path of the electron beam between the plates. The fields deflect the electron beam.

(a) Explain how the electrons that form the beam are released from the cathode by thermionic emission. **(2 marks)**

When a metal is heated, the electrons inside it gain kinetic energy, and some are released from the surface of the metal. These can then be accelerated by an applied electric field.

(b) The electrons are then accelerated through a potential difference of 2500 V.
 (i) Calculate the kinetic energy of an electron as it enters the region between the deflecting plates. Give your answer in electronvolts and joules. **(2 marks)**

Kinetic energy = 2500 eV = 2500 × 1.60 × 10⁻¹⁹
$$= 4.0 \times 10^{-16}\,J$$

 (ii) Calculate the speed of the electron in (ii). **(2 marks)**

$\frac{1}{2}mv^2 = 4.0 \times 10^{-16}$

so $v = \sqrt{(2 \times 4.0 \times 10^{-16}/9.11 \times 10^{-31})} = 3.0 \times 10^7\,m\,s^{-1}$

(c) A separate potential difference is connected across the two deflection plates so that the beam deflects upwards as shown. The electric field strength between the plates is 12 000 V m⁻¹.
 (i) State the polarity of the top plate. **(1 mark)**

Positive. It must attract the electrons.

 (ii) Calculate the force on an electron when it is between the plates. **(1 mark)**

$F = Ee = 12\,000 \times 1.6 \times 10^{-19} = 1.9 \times 10^{-15}\,N$

(d) A student uses the external coils in order to create a magnetic field through the tube. As the current is increased, the upward deflection decreases.
 (i) State the direction of the magnetic field in the tube. **(1 mark)**

Into the page, using Fleming's left-hand rule.

 (ii) Calculate the magnetic field strength that would make the beam travel horizontally through the tube. Assume that the electric field is unchanged. **(3 marks)**

The magnetic force and electric force on the electron must be equal and opposite:

$eE = Bev$

$B = \frac{E}{v} = \frac{12\,000}{3.0} \times 10^7 = 4.0 \times 10^{-4}\,T$

You have probably seen a deflection tube in class, but even if you have not you can work out how it works from the description given at the start of the question. Look at the photograph and identify the parts referred to in the question:

- vacuum tube
- electron gun
- deflector plates
- coils.

You are asked to give the answer in both eV and J. An electronvolt is the energy transferred when a charge e is moved through a p.d. of 1 V. Electrons have charge e and here they are accelerated through a p.d. of 2500 V, so they must each gain 2500 eV of energy.

In order to calculate the speed of the electron in m s⁻¹ you must use kinetic energy in joules. This is because both m s⁻¹ and J are S.I. units and eV is not.

When using Fleming's left-hand rule, remember that the second finger points in the direction of **convential current**, + to −. Here we have a beam of electrons travelling right to left, so they carry conventional current from left to right.

Notice how this answer is set out. You might consider calculating an expression for the magnetic force and setting it equal to the electric force from part (c)(ii), and then solving for B. This is perfectly acceptable, but the algebraic approach used here is simpler!

Don't forget the unit.

Specific heat capacity

Energy must be supplied to increase the temperature of a material. The amount of energy needed for a given temperature increase depends on the material and its mass.

Specific heat capacity

Specific heat capacity c is defined as the energy needed to raise the temperature of 1 kg of material by 1 K.

$$c = \frac{\Delta E}{m\Delta\theta}$$

where $\Delta\theta$ is the increase in temperature.

Remember that a 1 K change in temperature is the same as a 1°C change in temperature.

S.I. unit: $J\,kg^{-1}K^{-1}$

Typical values of specific heat capacity

Material	SHC / $J\,kg^{-1}K^{-1}$
air	1005
water (liquid)	4200
concrete	880
aluminium	897
copper	385
iron	449

The very high value for the specific heat capacity of water makes it ideal as a coolant – it can absorb a lot of energy for a relatively small increase in its temperature.

Practical skills   **Measuring the specific heat capacity of a metal**

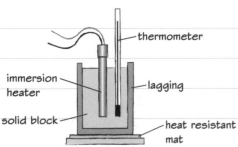

- thermometer
- immersion heater
- lagging
- solid block
- heat resistant mat

Apparatus for measuring the specific heat capacity of an aluminium block

$\Delta E = IVt$, the energy supplied by the immersion heater

$$c = \frac{IVt}{m\Delta\theta}$$

Measurements:

- current – use an ammeter
- p.d. across immersion heater – use a voltmeter
- time of heating – use a stop clock
- temperature change – use a thermometer.

1 Insulate the block to prevent heat transfer to or from the surroundings.

2 Do not start the clock immediately – some energy is needed to heat the heater.

Worked example

An experiment to measure the specific heat capacity of aluminium like the one shown left yielded the following results:

$I = 2.0\,A$, $V = 12.0\,V$, $t = 90\,s$, $m = 500\,g$, $\Delta\theta = 4.4°C$

(a) Calculate the specific heat capacity of aluminium. **(2 marks)**

$$c = \frac{IVt}{m\Delta\theta} = \frac{2.0 \times 12.0 \times 90}{0.50 \times 4.4} = 980\,J\,kg^{-1}K$$

(b) Compare your result from (a) with the expected value in the table and suggest a reason for the discrepancy. **(3 marks)**

The result of $980\,J\,kg^{-1}K^{-1}$ is about 10% larger than the expected value of about $897\,J\,kg^{-1}K^{-1}$. This is probably because some of the energy supplied was transferred to the environment. Some of the energy supplied will have been used to heat the heater itself rather than the block.

Now try this

1. Calculate the time it would take to raise the temperature of 800 g of water from 20°C to 80°C using a 50 W immersion heater. **(3 marks)**

2. An electric shower has a heater of power 10 kW and it supplies water at a rate of 7.0 litres per minute. 1 litre of water has a mass of 1 kg. The water entering the shower has a temperature of 20°C. Calculate the temperature of the water coming out of the shower. **(5 marks)**

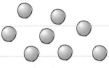
Latent heats

When a substance changes state, bonds form or are broken. This releases or requires energy.

Solids, liquids and gases

energy supplied →
← energy transferred away

energy supplied →
← energy transferred away

solid – particles bonded together in fixed positions

liquid – bonds breaking and reforming particles can flow past one another

gas – bonds broken, particles in random motion

When energy is supplied to a material by heating:
- the temperature of the material may increase with no change of state
- the state of the material may alter with no change in temperature.

Practical skills — Measuring the specific latent heat of vaporisation

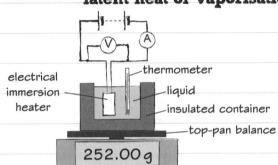

electrical immersion heater
thermometer
liquid
insulated container
top-pan balance

252.00 g

Measurements:
- t = time of heating
- Δm = reduction in mass
- I = current
- V = p.d. across heater

Calculation:

$$L = \frac{\Delta E}{\Delta m} = \frac{IVt}{\Delta m}$$

As heat is supplied the liquid changes to a gas and the mass falls. The thermometer is used to ensure that readings are taken only at the boiling point.

Worked example

A kettle contains 0.80 kg of water at 25°C. Calculate the minimum amount of energy needed to:

(a) Raise the temperature of the water to 100°C. **(2 marks)**

To raise temperature with no change of state we use the specific heat capacity.

$\Delta E = mc\Delta\theta = 0.80 \times 4200 \times 75$
$= 250\,000\,J$ (2 s.f.)

(b) Change the state of all of the water from a liquid to a gas at 100°C. **(2 marks)**

To change the state of the water at its boiling point we use the specific latent heat of vaporisation.

$\Delta E = L\Delta m = 2\,260\,000 \times 0.80$
$= 1\,800\,000\,J$ (2 s.f.)

Phase changes and latent heat

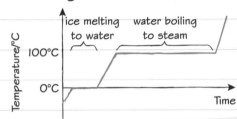

Temperature/°C
ice melting to water   water boiling to steam
100°C
0°C
Time

Temperature changes as ice below 0°C is supplied with energy at a constant rate until it is all converted to steam at above 100°C.

The energy supplied to change the state of the material (with no change in temperature) is called the **latent heat**. The **specific latent heat** L is the energy required to change the state of one kilogram of a substance at a constant temperature.

$$\Delta E = L\Delta m$$

where ΔE is the energy supplied and Δm is the mass of material changing state.

L_f = specific latent heat of fusion – melting/freezing
L_v = specific latent heat of vaporisation – boiling/condensing
S.I. unit = $J\,kg^{-1}$
For water, $L_f = 336\,kJ\,kg^{-1}$, $L_v = 2260\,kJ\,kg^{-1}$

Now try this

A student carries out an experiment to measure the latent heat of vaporisation of water using apparatus like that shown above. Her results are:

$\Delta m = 0.0060\,kg$, $I = 2.5\,A$, $V = 12.0\,V$, $t = 8.0$ minutes.

(a) Calculate a value for the latent heat of vaporisation of water. **(3 marks)**

(b) Compare your calculated value with the accepted value of $2260\,kJ\,kg^{-1}$ and comment on what might cause any difference. **(2 marks)**

Pressure and volume of an ideal gas

When an ideal gas is compressed or expands at constant temperature, its pressure and volume vary in inverse proportion.

Practical skills **An experiment to investigate pressure and volume of an ideal gas**

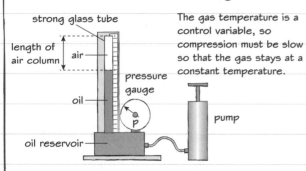

strong glass tube
length of air column
air
oil
pressure gauge
P
pump
oil reservoir

The gas temperature is a control variable, so compression must be slow so that the gas stays at a constant temperature.

Measurements:
- length of air column: proportional to the volume of the gas
- pressure: read from the Bourdon pressure gauge.

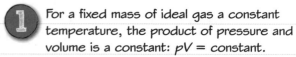
Maths skills **Boyle's law**

The results of the experiment show that:

① For a fixed mass of ideal gas a constant temperature, the product of pressure and volume is a constant: pV = constant.

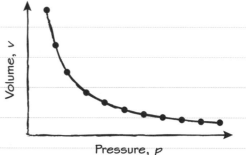

Volume, v

Pressure, p

② Pressure is inversely proportional to volume for a fixed mass of ideal gas a constant temperature: $p \propto \frac{1}{V}$.

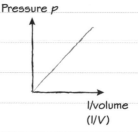

Pressure p

1/volume ($1/V$)

In other words, $p_2 V_2 = p_1 V_1$ where the subscripts 1 and 2 represent the initial and final states respectively.

This is **Boyle's law**.

Worked example

1 Why was it important to keep the temperature constant in the experiment described above? **(2 marks)**

If the temperature changed the velocities of the gas molecules would change and this would affect the pressure. For the experiment to be a fair test, only one independent variable, volume, must vary, so temperature must be constant.

2 A container of volume 0.050 m³ contains an ideal gas at a pressure of 5.0×10^5 Pa. The gas escapes from the container and enters the atmosphere at a pressure of 1.0×10^5 Pa. Calculate the volume of the escaped gas, assuming there is no change in temperature. **(3 marks)**

Using Boyle's law: $p_2 V_2 = p_1 V_1$

$$V_2 = \frac{p_1 V_1}{p_2} = \frac{5.0 \times 10^5 \times 0.05}{1.0 \times 10^5} = 0.20 \, m^3$$

For a reminder about Kinetic theory of gas see page 98

Now try this

1 An ideal gas is compressed at constant temperature from a volume of 0.060 m³ to 0.020 m³. Its initial pressure was 1.2×10^5 Pa.
   (a) Calculate its final pressure. **(3 marks)**
   (b) Explain how the final pressure would be affected if the gas were to heat up during compression. Explain. **(3 marks)**
   (c) Explain how the final pressure would be affected if there was a very slight leak in the container and some gas escaped during compression (assume there is no change in temperature). **(2 marks)**

2 Suggest in terms of molecular motion why compressing a gas at constant temperature increases its pressure. **(2 marks)**

Absolute zero

At the temperature absolute zero, the particles of an ideal gas would have no kinetic energy at all.

Gas pressure and temperature

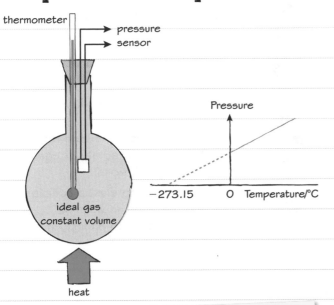

Increasing the temperature of an ideal gas increases its pressure. If a graph of pressure against temperature is plotted and the line is extended backwards, it intercepts the temperature axis at about −273°C, whatever gas is used and however much gas is used. This suggests that all ideal gases exert no pressure at this temperature.

Gas volume and temperature

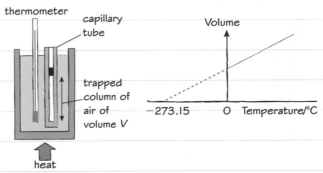

Increasing the temperature of an ideal gas increases its volume. If a graph of volume against temperature is plotted and the line is extended backwards, it intercepts the temperature axis at about −273°C, whatever gas is used and however much gas is used. This suggests that all ideal gases occupy zero volume at this temperature.

The thermodynamic temperature scale

The thermodynamic, Kelvin, scale starts at absolute zero, 0 K, defined as −273.15°C, and has divisions the same size as those in the Celsius scale.

temperature in Celsius

$$T = \theta + 273.15$$

temperature in kelvin

If appropriate we often round 273.15 to 273.

On the Kelvin scale, pressure and volume of an ideal gas are both directly proportional to temperature, $p \propto T$ and $V \propto T$.

Worked example

1 Convert the following temperatures to the Kelvin scale:

(a) a room temperature of about 21°C   **(1 mark)**

$T = 21 + 273 = 294\,K$

(b) the boiling point of water at 100°C.   **(1 mark)**

$T = 100 + 273.15 = 373.15\,K$

2 Convert the following temperatures to the Celsius scale:

(a) the temperature of empty space, 2.7 K   **(1 mark)**

$\theta = 2.7 - 273.15 = -270.5°C$

(b) The temperature at the surface of the Sun, 5800 K.   **(1 mark)**

$\theta = 5800 - 273 = 5527°C$

Now try this

1 The normal body temperature of a human being is about 37°C. Express this on the thermodynamic, Kelvin scale.   **(1 mark)**

2 The temperature of a constant volume of an ideal gas at a pressure of 1.01×10^5 Pa is increased from 50°C to 120°C.

(a) Express both temperatures in Kelvin.   **(2 marks)**

(b) Express the increase in temperature in both Celsius and Kelvin.   **(2 marks)**

(c) Calculate the final pressure of the gas.   **(4 marks)**

Kinetic theory

The behaviour of an ideal gas can be explained by a simple model.

The ideal gas equation

A constant mass of an ideal gas obeys three gas laws:

1 $p \propto \dfrac{1}{V}$ (at constant temperature)

2 $p \propto T$ (for constant volume)

3 $V \propto T$ (for constant pressure)

Therefore $pV \propto T$, or:

$$pV = \text{constant} \times T$$

The constant is proportional to the number of molecules N. It can be written as Nk, where k is the Boltzmann constant:

$$pV = NkT$$

Assumptions of the kinetic theory

1 Gases consist of a large number of identical molecules, so statistics can be applied.

2 The molecules are in rapid random motion.

3 The volume of each molecule is negligible.

4 The molecules do not exert long-range forces on one another, so affect each other only during collisions.

5 The molecules undergo elastic collisions with one another and the walls of their container, losing no energy, in instantaneous collisions.

Deriving the kinetic theory equation

Consider a molecule colliding with the wall of a container. The force it exerts on the wall = rate of change of momentum at the wall. The change of momentum in the x direction at this wall will be $-2mc_x$ where c_x is the x-component of the molecular velocity.

The molecule bounces back in the opposite direction and travels the length a of the container before bouncing off the far wall and returning. The time before the next collision of the molecule with the first wall is therefore $\dfrac{2a}{c_x}$

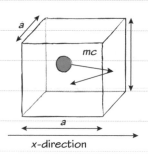

A single gas molecule colliding with one wall of a container.

x-direction

The average rate of change of momentum at that wall from this molecule alone is thus $-\dfrac{mc_x^2}{a}$

Pressure is force per unit area so the pressure contribution from one molecule is $\dfrac{mc_x^2}{a^3}$

For N molecules the pressure on the wall will be the sum of N contributions, taking the mean square speed $<c_x^2>$. This is calculated by squaring the individual speed of each molecule and finding the mean of the squares.

$p = \dfrac{Nm<c_x^2>}{V}$ where V is the volume = a^3 of the container.

But this is considering only the speed in the x-direction. For all motion, $<c_x^2> = \frac{1}{3}<c^2>$ because $<c^2> = <c_x^2> + <c_y^2> + <c_z^2>$, and the motion is random so $<c_x^2> = <c_y^2> = <c_z^2>$

$pV = \frac{1}{3}Nm<c^2>$

Worked example

Show that $\dfrac{pV}{T} = $ constant for a constant mass of an ideal gas. **(3 marks)**

$pV = NkT$

$\dfrac{pV}{T} = Nk$

For a fixed mass of gas N is constant.

k is also a constant.

Therefore $\dfrac{pV}{T} = $ constant.

Now try this

A flask of volume 0.0040 m³ contains air at pressure 2.0×10^5 Pa and temperature 50 °C. The flask is heated to 90°C. (Avogadro number = 6.02×10^{23}, Boltzmann constant $k = 1.38 \times 10^{-23}$ J K⁻¹)

(a) Calculate the number of air molecules inside the flask. **(3 marks)**

(b) Calculate the final pressure in the flask. **(2 marks)**

Particles and energy

When matter is heated, the random thermal kinetic energies of particles increase.

Internal energy

The particles – molecules, atoms or ions – that make up matter are always in motion.

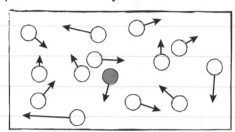

Each molecule in matter has:
- kinetic energy due to motion
- potential energy due to its interaction with other molecules.

The **internal energy** U of a body is the sum of the random kinetic energies and potential energies of all molecules in the body.

The internal energy can change if:
- the body is heated or cooled
- work is done on or by the body.

Worked example

Air consists mainly of oxygen with a molar mass 32 g and nitrogen with a molar mass 28 g. (Avogadro number = 6.02×10^{23}, Boltzmann constant $k = 1.38 \times 10^{-23}$ J K^{-1})

(a) Explain why the mean square speed of oxygen molecules will be lower than the mean square speed of nitrogen molecules in a room at 20°C. **(3 marks)**

$<KE> = \frac{1}{2}m<c^2> = \frac{3}{2}kT$, so at the same temperature both types of molecule have the same mean kinetic energy. However, the oxygen molecules have more mass so their mean velocities will be smaller.

(b) Calculate the root mean square, $\sqrt{<c^2>}$, speed of a nitrogen molecule in this room. **(3 marks)**

$\frac{1}{2}m<c^2> = \frac{3}{2}kT$

so $<c^2> = 3kT/m$ and

$\sqrt{<c^2>} = \sqrt{(3kT/m)}$

$= \sqrt{\dfrac{(3 \times 1.38 \times 10^{-23} \times 293 \times 6.02 \times 10^{23})}{28 \times 10^{-3}}}$

$= 511 \, \text{m s}^{-1}$

The mass of a mole of nitrogen molecules is 0.028 kg, so the mass of one molecule is this value divided by the number of molecules in a mole, the Avogadro number.

Internal energy of an ideal gas

The molecules in an ideal gas do not exert forces on one another except in collisions. This means that they have no potential energy.

Therefore the internal energy of an ideal gas is the sum of kinetic energies of all molecules in the gas.

The mean kinetic energy ($<KE>$) of a molecule in an ideal gas is given by the equation:

$$<KE> = \frac{1}{2}m<c^2>$$

where all molecules have mass m and $<c^2>$ is the mean square speed, or mean square velocity.

Energy and temperature

Combining the ideal gas equation and the kinetic theory equation:

$pV = NkT$

$pV = \frac{1}{3}Nm<c^2>$

Therefore $\frac{1}{3}Nm<c^2> = NkT$

Since the mean kinetic energy $<KE> = \frac{1}{2}m<c^2>$ we can use this expression to get

$$<KE> = \frac{1}{2}m<c^2> = \frac{3}{2}kT$$

The mean kinetic energy of a molecule of an ideal gas is directly proportional to the thermodynamic temperature, and at absolute zero (0 K) the molecules have no kinetic energy. This explains why ideal gases would exert zero pressure at 0 K – the molecules have stopped moving.

Now try this

1. Calculate the root mean square speed of oxygen molecules in the room described in the worked example. **(3 marks)**

2. Suggest a reason why the rate of a chemical reaction increases when the mixture is heated. **(2 marks)**

3. Five molecules in a gas have the following speeds in m s^{-1}: 310, 325, 362, 348, 360
   (a) Calculate the mean speed. **(2 marks)**
   (b) Calculate the mean square speed. **(2 marks)**
   (c) Calculate the root mean square speed. **(2 marks)**

$<c^2>$ is calculated by squaring the individual speed of each molecule and finding the mean of the squares. Be careful – it is not the square of the mean speed!

Black body radiation

Everything emits thermal radiation, but an ideal 'black body' radiates a distinctive spectrum.

Black body radiation

When heat is supplied to an object its particles vibrate more vigorously and it emits electromagnetic radiation more intensely. The peak wavelength of the radiation also shifts towards shorter wavelengths – think about an electric grill heating up and changing from black through dull red to bright orange.

Cool objects emit mainly in the infrared. As they get hotter they become 'red hot' and then white hot once most of the radiant energy emitted is in the visible spectrum.

An ideal black body radiator is one that absorbs and emits all wavelengths.

The black body radiation spectrum

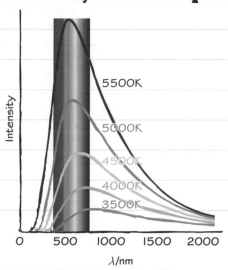

Emission from a black body at various temperatures. As temperature increases more energy is radiated and the peak of radiated intensity λ_{max} moves to shorter wavelengths.

Wien's law

As T increases λ_{max} becomes smaller, but the product of λ_{max} and T is constant.
This is **Wien's law:**

$$\lambda_{max} \propto \frac{1}{T}$$

$$\lambda_{max}T = 2.898 \times 10^{-3}\,\text{m K}$$

Wien's law enables astronomers to work out the surface temperature of a star by assuming it is a black body and using the value of λ_{max} from its spectrum.

The peak wavelength from the red giant star Betelgeuse is about 900 nm (infrared).

$$T = \frac{2.90 \times 10^{-3}}{900 \times 10^{-9}} = 3200\,\text{K}$$

The Stefan Boltzmann law

The **Stefan–Boltzmann law** states that as the temperature of a black body increases, its luminosity L, the total power it radiates, is directly proportional to its surface area A and to the fourth power of its temperature T, $L \propto AT^4$.

$$L = \sigma AT^4$$

where $\sigma = 5.67 \times 10^{-8}\,\text{W m}^{-2}\,\text{K}^{-4}$ is the Stefan–Boltzmann constant.

Worked example

The human body has a temperature of about 37°C.

(a) Calculate the wavelength at the peak of the emitted spectrum, λ_{max}. Assume that the spectrum of radiation emitted from the surface of a human is similar to the spectrum emitted by a black body. State into which part of the electromagnetic spectrum most of the energy radiated by a human falls. **(2 marks)**

$$\lambda_{max}T = 2.90 \times 10^{-3}$$
$$\lambda_{max} = \frac{2.90 \times 10^{-3}}{310} = 9.4 \times 10^{-6}\,\text{m},$$

which is in the infrared.

(b) Assume that the surface area of a human is about $1.8\,\text{m}^2$. Calculate the total power radiated by the human body. **(2 marks)**

$$L = \sigma AT^4$$
$$L = 5.67 \times 10^{-8} \times 1.8 \times 330^4 = 1200\,\text{W}$$

Don't forget to convert temperatures in Celsius to the Kelvin scale!

Now try this

The Sun has a surface temperature of 5800 K and a diameter of 1.4×10^6 km.
(a) Calculate the wavelength at the peak of its emitted spectrum and state in which part of the electromagnetic spectrum this wavelength falls. **(3 marks)**
(b) Calculate the total energy radiated by the Sun. **(2 marks)**

Standard candles

In order to work out how far away something is astronomers need to know how bright it is.

The inverse square law

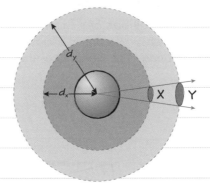

A star radiating uniformly in all directions.

Area of the sphere over which the total power of the star is distributed at distance $d = 4\pi d^2$.

$$\text{Intensity} = \frac{\text{power}}{\text{area}}, \text{W m}^{-2}$$

Space is a vacuum, so no radiation is absorbed. However, it is distributed over a wider area with distance from the source, so intensity is lower at Y than at X.

Luminosity L = total power output from source (W)

Intensity I at distance $d = \dfrac{\text{luminosity}}{\text{surface area of sphere of radius } d}$

$$I = \frac{L}{4\pi d^2}$$

Comparing intensities at d_X and d_Y: the areas of X and Y are related by their distance d_X and d_Y from the source. The area increases as the square of the distance, so:

$$\frac{I_X}{I_Y} = \frac{d_Y^2}{d_X^2}$$

Using the inverse square law to calculate distance

If an astronomical object has luminosity L, then the intensity of radiation reaching the Earth will be:

$$I = \frac{L}{4\pi d^2}$$

If we know the luminosity of the object, we can measure I and thus calculate the distance d:

$$d = \sqrt{\left(\frac{4\pi I}{L}\right)}$$

Standard candles

Astronomers call objects whose luminosity is known **standard candles**. Several types are commonly used, but two of the most important are:

- cepheid variable stars
- type Ia supernovae.

In both cases astronomers can calculate their luminosity by watching how the light received from them (their light curve) varies with time.

Worked example

The intensity of solar radiation received at the edge of the Earth's atmosphere is about $1400\,\text{W m}^{-2}$ and the distance from the Earth to the Sun is about $1.5 \times 10^8\,\text{km}$.

Calculate the luminosity of the Sun. State any assumptions you have made. **(4 marks)**

$$L = 4\pi d^2 I = 4 \times \pi \times (1.5 \times 10^{11})^2 \times 1400$$
$$= 4.0 \times 10^{26}\,\text{W}$$

We must assume that the Sun radiates uniformly in all directions so that its radiation obeys an inverse square law. We also assume that there is nothing between the Earth and Sun that absorbs or scatters the radiation.

A light year is the distance light travels in one year at $c = 3.00 \times 10^8\,\text{m s}^{-1}$.

Now try this

The star Betelgeuse has a luminosity of $5.6 \times 10^{31}\,\text{W}$ and is 643 light years from Earth. Calculate the intensity of radiation from Betelgeuse when it reaches the Earth. **(3 marks)**

Trigonometric parallax

The movement of the Earth around the Sun provides astronomers with another way to calculate the distance to nearby astronomical objects.

Using parallax angles to calculate distances

As the Earth orbits the Sun, the apparent position of a 'nearby' star changes against the background of very distant stars, in the same way that, when you are in a moving train, the trees near the track seem to be moving against the distant background. The closer the star the greater the change, or **parallax**.

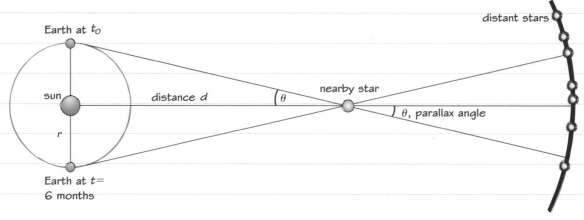

We know the radius of the Earth's orbit around the Sun and we can use telescopes to measure the parallax angle, so we can use trigonometry to calculate the distance of the star.

$$\frac{r}{d} = \tan \theta$$

 Maths skills **Small-angle approximations**

Try working out trigonometric functions for very small angles in radians. You will find that as $\theta \to 0$: $\sin \theta \to \theta$; $\tan \theta \to \theta$; $\cos \theta \to 1$

Because stars are so very far away, distance d is always very much larger than the radius of the Earth's orbit, making the parallax angles so small that we can use the small-angle approximation, $\tan \theta = \theta$ (in radians).

$$\frac{r}{d} = \theta$$

$$\text{distance } d = \frac{r}{\theta}$$

Limitations

For more distant stars the parallax angle is too small to be measured, so this method is only useful out to about 10^{13} m.

The method can be extended by using a space telescope. This increases the precision of parallax measurements because the light reaching the telescope does not have to pass through the Earth's atmosphere.

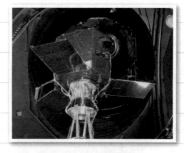

The Hipparcos spacecraft can measure the parallax angles of stars out to almost 10^{14} m from Earth.

Worked example

1 Our next-nearest star, Proxima Centauri, has a stellar parallax of 3.7×10^{-6} rad. The radius of the Earth's orbit around the Sun is 1.5×10^{11} m.

Calculate the distance to Proxima Centauri.
(2 marks)

$$d = \frac{r}{\theta}$$

$$\text{Distance to Proxima Centauri} = \frac{1.5 \times 10^{11}}{3.7 \times 10^{-6}}$$

$$= 4.1 \times 10^{16} \text{ m}$$

2 Explain why the 'background of very distant stars' does not appear to move as the Earth orbits the Sun.
(2 marks)

These stars are so far away that their parallax angles are too small to be detected.
This makes them appear as a fixed background against which closer stars seem to move.

Now try this

1 Calculate the distance of a star with a parallax angle of 10^{-7} radians.
(2 marks)

2 Calculate the distance of a star with a parallax angle of 3×10^{-5}°.
(3 marks)

The Hertzsprung–Russell diagram

Stars can be classified by plotting a graph of luminosity against temperature.

The Hertzsprung–Russell (HR) diagram

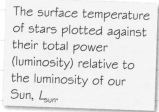

The surface temperature of stars plotted against their total power (luminosity) relative to the luminosity of our Sun, L_{sun}.

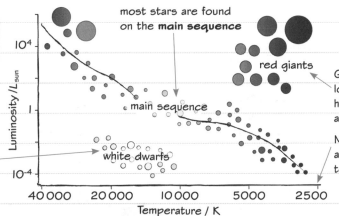

most stars are found on the **main sequence**

red giants

Giants and supergiants have low surface temperatures and high luminosity, so must have a large surface area.

White dwarfs have high surface temperature and low luminosity, so must have a small surface area.

Note that the temperature axis is reversed – higher temperatures are to the left.

When the luminosity of stars is plotted against their surface temperature, several distinct classes of star emerge.

The temperature scale in the HR diagram runs from high to low because originally the graph was plotted using the peak emission wavelength, λ_{max}. This was later found to be proportional to $1/T$ as described by Wien's law – see page 100. The conclusions about surface area are based on the Stefan–Boltzmann law, $L = \sigma A T^4$, also described on page 100.

Main-sequence stars

Stars in the main sequence are fusing hydrogen nuclei into helium nuclei in their cores. In the HR diagram the main sequence has star mass increasing from bottom right to top left.

When a star begins to run out of hydrogen for fusion, it changes type on the HR diagram. The more massive the star, the less time it spends on the main sequence. Stars of average mass, such as our Sun, remain on the main sequence for billions of years. The massive blue giants at the top left of the main sequence may only spend a few million years on the main sequence, as they exhaust the supply of hydrogen in their core much faster than less massive stars.

Worked example

Use the HR diagram above to calculate the relative difference in luminosity between a low-mass and a high-mass main-sequence star. **(2 marks)**

Low-mass star has a luminosity $\approx 10^{-4} \times L_{Sun}$

High-mass star has a luminosity $\approx 10^4 \times L_{Sun}$

So relative difference $\approx \dfrac{10^{-4} \times L_{Sun}}{10^4 \times L_{Sun}} = 10^8$

🖩 Maths skills

$0.001 = 10^{-3}$	$\log_{10}(0.001) = -3$
$1 = 10^0$	$\log_{10}(1) = 0$
$1000 = 10^3$	$\log_{10}(1000) = 3$
$1\,000\,000 = 10^6$	$\log_{10}(1\,000\,000) = 6$

'$\log_{10}$' is often simply written '$\log$'.

Do not confuse this with $\log_e$ or $\ln$, which represent natural (base e) logs.

Now try this

(a) Sketch the HR diagram showing its four main regions and mark the position of the Sun and another main-sequence star that has a lower surface temperature than the Sun. **(6 marks)**

(b) Mark in the position of:
   (i) Betelgeuse, a red supergiant star **(1 mark)**
   (ii) Aldebaran, a red giant star **(1 mark)**
   (iii) Sirius B, a white dwarf star **(1 mark)**

103

Stellar life cycles and the Hertzsprung–Russell diagram

The fate of a star depends on its mass.

Stellar evolution

Stars are formed from collapsing gas clouds. As the protostar accumulates more gas and dust, it increases in density and temperature. When the protostar's collapse under its own gravity makes it dense and hot enough, hydrogen nuclei are forced together and begin to fuse into helium nuclei, releasing large amounts of energy. What happens next depends on the balance between the outward pressure, due to the hot particles and the radiation emitted, and the gravitational attraction pulling the particles inward. The evolution of a star therefore depends on its mass.

Stars release energy from nuclear fusion reactions in their cores. When the fuel begins to run out, gravity causes the star to collapse. Low-mass stars, such as our Sun, end up as white dwarfs. Higher mass stars undergo a supernova explosion and their cores collapse to form neutron stars or even black holes.

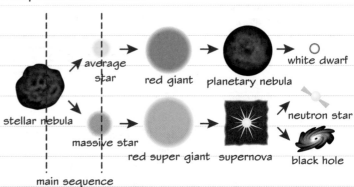

The life of our Sun on the HR diagram

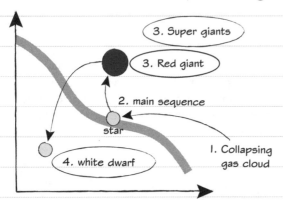

Our Sun was formed from a collapsing gas cloud. In a few billion years when most of the hydrogen in its core has been used up, changes will occur that cause the outer layers to expand, giving a larger, cooler **red giant**. Eventually the outer layers are ejected altogether, leaving a core in which no more fusion occurs, a **white dwarf**.

Red giants and white dwarfs

The red giants and red supergiants located at the top right of the HR diagram are cool stars with a very large surface area. Hence they have a high luminosity despite their low surface temperature.

White dwarfs are located towards the bottom left of the HR diagram. Although they have a high surface temperature, their surface area is relatively small. Hence they have a low luminosity despite their high surface temperature.

As a massive star, more than about ten times the mass of our Sun, reaches the end of its time on the main sequence it becomes a **red supergiant**, rather than a red giant, and evolves differently as shown above.

Now try this

1 Larger stars have shorter lifetimes – suggest and explain a reason for this. **(3 marks)**

2 Sketch the HR diagram and mark the life cycle of a massive star from the top left of the main sequence, starting with its formation up to the point before it goes supernova. **(4 marks)**

3 Suggest why Jupiter, a gas giant planet, did not become a star. **(3 marks)**

The Doppler effect

You can hear the sound of a car engine change frequency as it passes you; this is an example of the Doppler effect.

Doppler shifts

 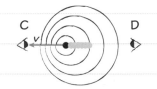

source of waves is stationary relative to observer

source of waves is moving relative to observer

Observers A and B are at rest with respect to the wave source. They both receive the same wavelength and frequency as the source emits.

Observer C and the source are getting closer together. The wave frequency is increased and the wavelength reduced compared with waves when the source and observer remain stationary with respect to each other. This is a **Doppler shift** to higher frequency.

Observer D and the source are becoming further apart. The frequency is reduced and the wavelength increased compared with waves when the source and observer remain stationary with respect to each other – a Doppler shift to lower frequency.

Doppler shift for electromagnetic waves

Atoms in a hot gas absorb light at certain frequencies characteristic of that element. It was observed that these **spectral lines** are shifted towards the red end of the spectrum in light observed from distant galaxies compared with the same lines recorded in the laboratory.

This **red shift** suggests that those galaxies and the Earth must be moving apart, Doppler-shifting the wavelength of the light we see.

The speed of EM waves, c, is constant regardless of the motion of the source or observer. For EM waves, as long as the relative velocity of the objects $v \propto c$, the Doppler shifts are given by:

$$z = \frac{\Delta\lambda}{\lambda} \approx \frac{\Delta f}{f} \approx \frac{v}{c}$$

where λ and f are the wavelength and frequency when the source is stationary.

Hubble's law

The red shifts z of distant galaxies were found to be directly proportional to their distance d. This means that relative recession velocities are also directly proportional to distance, $v \propto d$ or $v = H_0 d$ where H_0 is the Hubble constant (S.I. unit: s^{-1})

Cosmological red shifts

The red shifts observed for distant galaxies can be explained if the Universe itself is expanding. The expansion of space increases the distance between galaxies and 'stretches' the wavelength of electromagnetic waves.

The red shift z can be used to find the relative velocity of the receding galaxies: $z \propto \frac{v}{c}$, so $v \propto zc$.

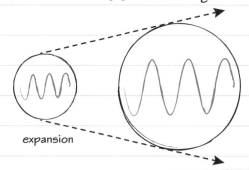

expansion

As space expands, it increases the wavelength of the radiation it contains.

Worked example

Hydrogen has a spectral line at a wavelength of 656 nm. When this spectral line is observed in light from another galaxy its wavelength has shifted to 664 nm. Calculate the red shift z and the recession velocity of the galaxy containing this star. ($c = 3.00 \times 10^8\,\mathrm{m\,s^{-1}}$) **(4 marks)**

$\Delta\lambda = \lambda' - \lambda = 664 - 656 = 8\,\mathrm{nm}$

$z = \dfrac{\Delta\lambda}{\lambda} = \dfrac{8}{656} = 0.012 = \dfrac{v}{c}$

Therefore $v = 0.012 \times c = 3.7 \times 10^6\,\mathrm{m\,s^{-1}}$, moving away.

Now try this

1 Sodium vapour emits a spectrum containing a prominent spectral line at a wavelength of 589 nm. When this spectral line is received from a star in a distant galaxy its wavelength has shifted to 595 nm. Calculate the red shift z and the recession velocity of the galaxy containing this star. ($c = 3.00 \times 10^8\,\mathrm{m\,s^{-1}}$) **(4 marks)**

2 The Hubble constant has a value of about $2.0 \times 10^{-18}\,\mathrm{s^{-1}}$. How far away is a galaxy with a red shift of $z = 0.050$? **(4 marks)**

Cosmology

Cosmology is the study of the Universe as a whole: its origin, evolution, structure and fate.

The expanding Universe

Hubble's law ($v = H_0 d$) suggests that:

 All distant galaxies are moving away from us.

 Recession velocities are directly proportional to distance.

This result can be explained if the Universe as a whole is expanding.

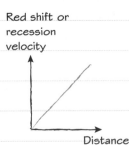

Red shift or recession velocity

Distance

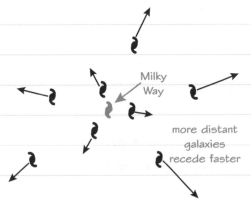

Milky Way

more distant galaxies recede faster

It might seem as if we are at the centre of this expansion, but in fact it would look the same seen from any galaxy – all the others would recede and obey Hubble's law.

The origin and age of the Universe

If galaxies are moving apart now they must have been closer together in the past. Tracing backwards, we find they would have all been present at a point about 13.7 billion years ago.

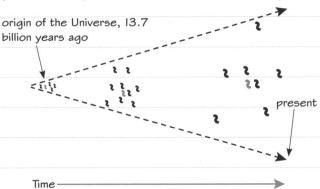

origin of the Universe, 13.7 billion years ago

present

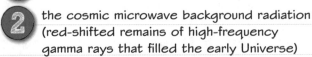

Time

This is the **Big Bang theory**: the Universe (space, time and matter) came into being in a hot Big Bang (explosion) 13.7 billion years ago, and the Universe has been expanding ever since.

Evidence to support the expanding Universe model:

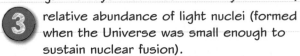

1 red shifts and Hubble's law

2 the cosmic microwave background radiation (red-shifted remains of high-frequency gamma rays that filled the early Universe)

3 relative abundance of light nuclei (formed when the Universe was small enough to sustain nuclear fusion).

The Hubble time

The age of the Universe can be estimated from the value of the Hubble constant. The larger H_0 is, the greater the rate of expansion and so the younger the Universe must be.

Hubble time $T = \dfrac{1}{H_0}$

If $H_0 \approx 2 \times 10^{-18} \, s^{-1}$ then $T = 5 \times 10^{17} \, s$

This is about 16 billion years. However, this method assumes that the recession velocities have been constant since the Big Bang, whereas in fact they have changed. Our most precise estimate of the age of the Universe comes from detailed studies of the background radiation and gives a value of about 13.7 billion years.

Dark matter and dark energy

The normal matter making up stars and galaxies is not enough mass to explain the motions of stars inside rotating galaxies. These unpredicted gravitational effects can be explained in terms of the effects of **dark matter**, so called because we cannot see it, we can only detect its presence through its effects on normal matter.

In addition, calculations about the expansion of the Universe suggest that dark energy may need to be taken into account. It now seems that normal matter makes up only 4% of the entire Universe!

Now try this

The fate of the Universe depends on the density of mass and energy within it. Explain why. **(3 marks)**

Exam skills 11

This exam-style question uses knowledge and skills you have already revised. Have a look at pages 96–100 for a reminder about gas pressure and the gas laws, Wien's law and the Stefan–Boltzmann law.

Worked example

1 A scuba diver breathes air from a gas cylinder. The volume of the cylinder is $0.0060\,\text{m}^3$ and the pressure of air inside the cylinder is $2.0 \times 10^7\,\text{Pa}$.

(a) Explain, in terms of molecules and Newton's laws of motion, how the air inside the cylinder exerts a pressure on its walls.   **(3 marks)**

The molecules are in rapid random motion. As they collide with the wall, the wall exerts a force on the molecules that changes their momentum. By Newton's third law, the molecules exert an equal but opposite force on the wall. By Newton's second law the average force on the wall is equal to the average rate of change of momentum of the molecules colliding with the wall. This creates a pressure on the wall because pressure is equal to force per unit area.

(b) Explain in terms of molecular motion why the pressure falls if the temperature of the gas is reduced.   **(3 marks)**

The mean kinetic energy per molecule is proportional to the absolute temperature T. As T falls, the molecules move more slowly, so collisions are less violent and less frequent. This reduces the average rate of change of momentum at the wall and reduces the pressure.

(c) Calculate the volume that would be occupied by the air contained in the cylinder if it all leaked out into the atmosphere, at pressure $1.0 \times 10^5\,\text{Pa}$, with no change of temperature.   **(3 marks)**

$pV = NkT$ so $V = \dfrac{NkT}{p}$

The only thing that changes is p, which falls by a factor of $(2.0 \times 10^7)/(1.0 \times 10^5) = 200$. The escaped air will occupy a volume $= 200 \times 0.0060 = 1.2\,\text{m}^3$.

2 The graph shows the radiation spectrum of the Sun. It is approximately a black-body spectrum. See the black line on the figure.

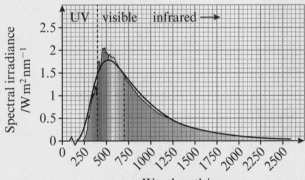

(a) Use the graph to show that the surface temperature of the Sun is about 6000 K and explain how you do this.   **(4 marks)**

The peak of the radiation curve occurs at about 520 nm.
Using Wien's law: $\lambda_p T = 2.9 \times 10^{-3}\,\text{m K}$
so $\lambda_p = 2.9 \times 10^{-3}/(520 \times 10^{-9}) = 5600\,\text{K}$ (about 6000 K)

(b) The radius of the Sun is 700 000 km. Use your answer to (d) to calculate the total power radiated by the Sun.   **(3 marks)**

From the Stefan–Boltzmann law:
$L = \sigma A T^4 = 4\pi r^2 \sigma T^4$
$ = 4\pi \times (7.0 \times 10^8)^2 \times 5.67 \times 10^{-8} \times 5600^4 = 3.4 \times 10^{26}\,\text{W}$

Read the question carefully – you need to use a particle model to answer this question and you must refer to Newton's laws of motion. It is worth pausing before answering this question to plan how you will structure your answer.

It would not be sufficient here simply to state that the molecules move more slowly. A complete answer must link to the way they cause pressure, through rate of change of momentum. There are two related effects – the molecules collide with the walls less frequently **and** less violently. Both reduce the average pressure on the container walls.

Note how the answer starts by quoting the relevant gas law: $pV = NkT$.

You are asked to 'use the graph'. What you have to do is read off the wavelength value at peak intensity. The best way to show how you have done this, and to actually get a value, is to draw a vertical line onto the graph using a ruler.

Mass and energy

Einstein showed that mass and energy are equivalent.

Einstein's mass–energy equation

Mass and energy are equivalent. This means that they can be interconverted. Their relationship can be calculated from Einstein's equation:

$$\Delta E = c^2 \Delta m$$

This equation implies that any process that releases energy must also result in a loss of mass. However, c^2 is a very large value, so in most processes the mass change is so small that it is negligible.

Mass–energy calculations

In a spontaneous reaction that releases energy, the mass of the products will be less than the mass of the reactants. For the difference in mass Δm for the reaction:

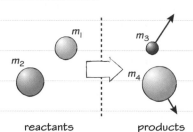

reactants   products

$(m_3 + m_4) < (m_1 + m_2)$

$\Delta m = (m_1 + m_2) - (m_3 + m_4)$

$\Delta E = c^2 \Delta m = c^2\{(m_1 + m_2) - (m_3 + m_4)\}$

Atomic mass units

In atomic and nuclear physics a non-S.I. unit called the atomic mass unit (u) is used.

$1u = \frac{1}{12} \times$ mass of a carbon-12 atom

$1u = 1.66 \times 10^{-27} kg$

This is approximately equal to the mass of a nucleon (proton or neutron).

The energy equivalent of 1 u

$$\Delta E = c^2 \Delta m = 9.00 \times 10^{16} \times 1.66 \times 10^{-27}$$
$$= 1.49 \times 10^{-10} J$$

This is equal to $\dfrac{1.49 \times 10^{-10}}{1.60 \times 10^{-19}} = 9.34 \times 10^8$
$$= 930\,MeV$$

Conservation of mass and energy

Look at the example of a spontaneous reaction shown on the left. The mass of the products is less than the mass of the reactants, so we cannot simply say that mass is conserved.

However, the products have more kinetic energy than the reactants, and energy is equivalent to mass. When we take this into account the total mass–energy before and after the reaction is the same.

Converting mass to energy

The world's largest nuclear explosion was the Tsar bomb (Big Ivan), exploded in 1961, which released 5000 times more energy than the Hiroshima atomic bomb, but converted only 2.3 kg of mass to energy.

Worked example

1 Calculate the maximum energy that could be released by combining 1.0 g of matter with 1.0 g of its antimatter counterpart.   **(2 marks)**

$$\Delta E = c^2 \Delta m = 9.00 \times 10^{16} \times 2.0 \times 10^{-3}$$
$$= 1.8 \times 10^{14} J$$

2 The mass of a uranium-238 atom is 238.0289 u. Calculate its rest energy in J and GeV.   **(3 marks)**

$$\Delta E = c^2 \Delta m$$
$$= 9.00 \times 10^{16} \times 238.0289 \times 1.66 \times 10^{-27}$$
$$= 3.56 \times 10^{-8} J$$

This is $\dfrac{3.56 \times 10^{-8}}{1.60 \times 10^{-19}} = 2.22 \times 10^{11}\,eV$
$$= 222\,GeV$$

Now try this

1 When a positron from a tracer used in a PET scan annihilates an electron inside a human body, a pair of gamma photons is emitted. These photons are detected and reveal where the tracer compound has gathered in the patient. Calculate the total energy released and hence the wavelength of the photons. ($m_e = 9.11 \times 10^{-31}$ kg, $h = 6.63 \times 10^{-34}$ J s)   **(4 marks)**

2 An iron-56 atom has a mass of 55.935 u. Calculate its rest energy in J and GeV.   **(3 marks)**

Nuclear binding energy

The energy released when a nucleus forms is called its binding energy.

Nuclear binding energy

The **binding energy** of a nucleus is equal to the energy needed to split the nucleus into individual nucleons and move them apart.

nucleus - all nucleons bound together

work must be done to separate the nucleons

separate nucleons

Energy must be supplied to separate the nucleons, so the total mass of the separated nucleons is greater than the mass of the nucleus:

Δm = mass deficit of nucleus = (total mass of separated nucleons) − (mass of nucleus)

Nuclear binding energy = $c^2 \Delta m$

Nuclear stability

The total binding energy of a nucleus increases as nuclei get bigger. To compare nuclear stability we use the **binding energy per nucleon**:

B.E. per nucleon = total B.E./A

where A = nucleon number

Variation of binding energy per nucleon with nucleon number

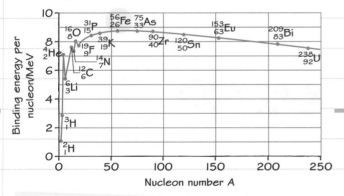

The most stable nucleus is iron-56. It has the largest binding energy per nucleon: that is, it requires the most energy per nucleon to separate its nucleons. Nuclei with greater stability tend to be more abundant in nature. Note the peaks for helium-4, carbon-12 and oxygen-16.

Calculating binding energy

Δm = (total mass of protons + total mass of neutrons) − (mass of nucleus)

Nuclear binding energy = $c^2 \Delta m$

Nuclear and atomic masses

The mass of a nucleus is equal to the atomic mass minus the mass of the orbiting electrons ($Z m_e$, where Z is the atomic number).

Worked example

Iron-56 has 26 protons and (56 − 26) = 30 neutrons in the nucleus. Use the data below to calculate the nuclear binding energy of the iron-56 nucleus and give your answer in J and MeV.
- atomic mass of iron-56 = 55.934939 u
- mass of an electron m_e = 0.000549 u
- mass of a proton m_p = 1.007276 u
- mass of a neutron m_n = 1.008665 u   **(5 marks)**

Nuclear mass
= (55.934938 − 26 × 0.000549) u
= 55.920664 u

Mass of separated nucleons
= (26 × 1.007276 + 30 × 1.008665) u
= 56.449126 u

Δm = (56.449126 − 55.920664) u
Δm = 0.528462 u

B.E.
= $9.00 \times 10^{16} \times 0.528462 \times 1.66 \times 10^{-27}$
= 7.90×10^{-11} J or 493 MeV (3 s.f.)

Maths skills When calculating nuclear binding energies or the energy released in nuclear reactions do not round off too early – the mass deficits usually depend on the final few decimal places.

Now try this

1 Look at the calculation of nuclear binding energy for iron-56 above. Calculate the binding energy per nucleon in MeV.   **(2 marks)**

2 The atomic mass of uranium-238 is 238.02891 u and its atomic number is 92.
   (a) Calculate the total nuclear binding energy for uranium-238 and express your answer in MeV.   **(6 marks)**
   (b) Calculate the binding energy per nucleon for uranium-238 in MeV.   **(2 marks)**

Nuclear fission

Heavy nuclei can be split to form lighter daughter nuclei and at the same time release a great amount of energy.

Binding energies in nuclear fission

The curve of binding energy (B.E.) per nucleon against nucleon number has a peak at iron-56. Therefore, if heavier nuclei split into two lighter nuclei with greater binding energy per nucleon, energy is released. This is **nuclear fission**.

$$\,^{1}_{0}n + \,^{235}_{92}U \rightarrow \,^{144}_{56}Ba + \,^{90}_{36}Kr + 2\,^{1}_{0}n$$

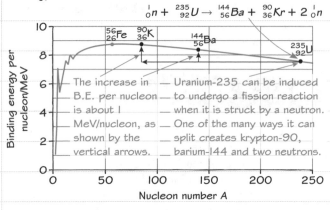

The increase in B.E. per nucleon is about 1 MeV/nucleon, as shown by the vertical arrows. — Uranium-235 can be induced to undergo a fission reaction when it is struck by a neutron. One of the many ways it can split creates krypton-90, barium-144 and two neutrons.

Calculating the energy released in nuclear fission

1 Calculate the mass of the reactants.

2 Calculate the mass of the products.

3 Calculate the change in mass Δm.

4 Use $\Delta E = c^2 \Delta m$.

Harnessing energy from nuclear fission

The isotope uranium-235 is naturally **fissile**. This means that when it absorbs a neutron it splits into daughter nuclei and releases a large amount of energy. This is induced nuclear fission.

Induced nuclear fission absorbs one neutron but usually emits two or more neutrons. This **chain reaction** is the key to extracting energy from fission. If these neutrons create further nuclear fission reactions, the reaction rate will grow, releasing energy. This release is uncontrolled in an atom bomb and controlled in a nuclear reactor.

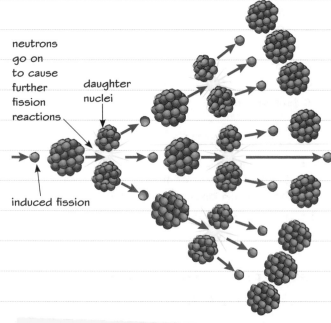

neutrons go on to cause further fission reactions

daughter nuclei

induced fission

A chain reaction in which the number of neutrons able to cause fission multiplies in each stage.

Worked example

Estimate the total energy that could be released from 1 kg of uranium-235 by the nuclear fission reaction yielding $^{90}_{36}K$ and $^{144}_{56}Ba$. Assume that 1 kg ^{235}U contains 2.56×10^{24} atoms. **(4 marks)**

$m_U = 235.043930\,u$ $m_{Kr} = 89.919517\,u$
$m_{Ba} = 143.922953\,u$ $m_n = 1.008665\,u$
$u = 1.66 \times 10^{-27}\,kg$

$\Delta m = m_U - (m_{Kr} + m_{Ba} + m_n)$
$= 235.043930\,u - (89.919517 + 143.922953 + 1.008665)\,u$
$= 0.192795\,u$
$= 0.192795 \times 1.66 \times 10^{-27}$
$= 3.200397 \times 10^{-28}\,kg$, per U atom

$\Delta E = c^2 \Delta m$
$= 9.00 \times 10^{16} \times 3.200397 \times 10^{-28}$
$= 2.8803573 \times 10^{-11}\,J$, per U atom

Energy released per kg U
$= 2.8803573 \times 10^{-11} \times 2.56 \times 10^{24}$
$= 7.4 \times 10^{13}\,J = 74\,000\,000\,MJ$

Now try this

1 A typical nuclear power station has an efficiency of 40% and an output of about 1 GW. How long could it run on the fission energy available from 1 kg of uranium-235? (Use the information from the worked example.) **(4 marks)**

2 Natural uranium consists of uranium-238 (99.3%) and a small amount of uranium-235 (0.7%). Uranium-238 is much more likely simply to absorb neutrons than to undergo induced nuclear fission. Explain why a runaway chain reaction cannot occur in a lump of natural uranium. **(2 marks)**

Nuclear fusion

Under extreme conditions, some light nuclei can fuse together to form heavier nuclei and to release a huge amount of energy.

Binding energies in nuclear fusion

Nuclear fusion is the process by which two light nuclei, e.g. isotopes of hydrogen, fuse to form a heavier nucleus, such as helium, and release a huge amount of energy.

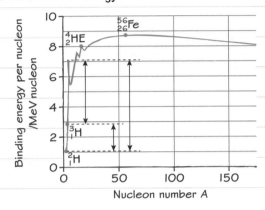

$$^2_1H + {}^3_1H \rightarrow {}^4_2He + {}^1_0n$$

Calculating the energy released in nuclear fusion

1 Calculate the mass of the reactants.

2 Calculate the mass of the products.

3 Calculate the change in mass Δm.

4 Use $\Delta E = c^2 \Delta m$.

> The nuclear fusion reaction illustrated on the left is the fusion of hydrogen-2, deuterium, and hydrogen-3, tritium, to form a helium nucleus, emitting one neutron. The energy released in the reaction becomes kinetic energy of the neutron and the helium-4 nucleus.

The challenge of fusion

Nuclear fusion occurs when **the strong nuclear force** overcomes the **electrostatic repulsion** between nuclei and binds them together. However, the strong nuclear force is very short range, so for fusion to happen the nuclei must be forced close together.

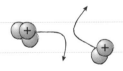

> The electrostatic repulsion between nuclei can only be overcome at extreme temperatures and densities.

These conditions can be created:

☑ in the core of a star

☑ in the centre of a nuclear explosion

☑ by focusing the world's most intense lasers onto a small sample of material

☑ in a magnetically confined plasma inside a fusion reactor.

Worked example

Use the data below to calculate the energy released by the nuclear fusion of deuterium, hydrogen-2, with tritium, hydrogen-3. **(4 marks)**

$m_n = 1.008665\,u$

atomic mass of deuterium $= 2.014102\,u$

atomic mass of tritium $= 3.016050\,u$

atomic mass of He-4 $= 4.002602\,u$

$\Delta m = (2.014102 + 3.016050) - (4.002602 + 1.008665) = 0.018885\,u$

$\Delta E = c^2 \Delta m$
$= 9.00 \times 10^{16} \times 0.018885 \times 1.66 \times 10^{-27}$
$= 2.82 \times 10^{-12}\,J = 17.6\,MeV$

> Both hydrogen atoms have one electron and the helium atom has two electrons, so we can ignore electrons because they will cancel in the calculation.

Now try this

1 A kilogram of deuterium and tritium mixed 1 : 1 contains enough reactants for 1.2×10^{26} reactions. Use the result from the worked example above to calculate the energy released per kilogram of reactants. **(2 marks)**

2 Explain why nuclear fusion reactions can only take place at high temperatures. **(3 marks)**

Background radiation

We live in a radioactive environment.

Sources of background radiation

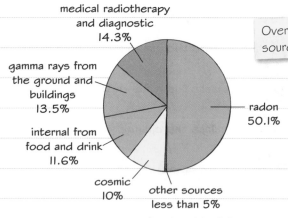

medical radiotherapy and diagnostic 14.3%

gamma rays from the ground and buildings 13.5%

internal from food and drink 11.6%

radon 50.1%

cosmic 10%   other sources less than 5%

> Overall mean radiation dose in the UK. Many of these sources are due to radioactive decay of unstable isotopes.

There are many sources of natural ionising radiation. The pie chart shows the contributions to our overall radiation dose in the UK, although the precise dose you receive from the background will depend on where you live and what you do.

Worked example

Suggest why airline pilots get a higher annual radiation dose from the background than people who work on the surface of the Earth. **(3 marks)**

The atmosphere shields people at ground level from much of the incoming cosmic radiation. Airline pilots spend long periods at high altitude with much less atmosphere above them. This reduces the shielding effect and increases their exposure to cosmic rays.

Taking background radiation into account in experiments

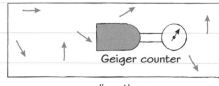

Geiger counter

radioactive source

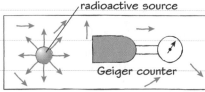

Geiger counter

1 Place the radiation detector, e.g. Geiger counter, in position without the source present.

2 Measure the count over a suitable period, e.g. 3 minutes, and repeat to obtain three readings.

3 Find the average count rate, counts per minute – this is the **background count**.

4 Repeat ❶ to ❸ but with the source present to find the average **total count**.

5 Subtract the average background count rate from the total count rate to obtain the average count rate from the source alone:

count rate (source) = total count rate – average background count rate

Now try this

1 Here are the results of an experiment with a radioactive source.

	Counts in 3 mins (1)	Counts in 3 mins (2)	Counts in 3 mins (3)
No source present	85	92	83
Source present	256	248	250

(a) Why do repeated measurements differ from one another? **(1 mark)**

(b) Calculate the average background count rate (cpm) in this experiment. **(3 marks)**

(c) Calculate the average total count rate (cpm) in this experiment. **(1 mark)**

(d) Calculate the average count rate (cpm) due to the source. **(2 marks)**

2 Radon-222 is a gas released from the radioactive decay of radium in rock, especially in granite. It can gather in poorly ventilated houses, especially basements, and mines. It is the second most common cause of lung cancer after smoking.

(a) Suggest why radon is likely to gather in basements or mines. **(2 marks)**

(b) Suggest why radon is particularly implicated in lung cancer as opposed to other sorts of cancer. **(1 mark)**

Alpha, beta and gamma radiation

Different types of ionising radiation can be distinguished by looking at their penetrating power, range and ionising ability in different substances.

Ionising radiation

Atoms of radioactive elements decay spontaneously and randomly. The resulting ionising radiation comes in three forms, which differ in their:

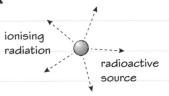

- **ionising power**: how many ions they can produce per unit distance in a particular material
- **penetrating power/range**: how far they can travel through various materials and what thickness of a particular material is needed to absorb them.

A single alpha, beta or gamma emission can ionise tens of thousands of air molecules as it travels, losing energy to each one until it is finally stopped.

Penetrating power

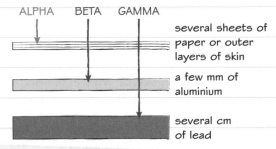

The absorbers needed to stop or very significantly reduce the three different types of ionising radiation.

Range in air

- alpha: 0.02–0.03 m
- beta: 1–2 m
- gamma: barely absorbed by air.

Summary of the nature and properties of ionising radiation

Type	Nature	Symbol	Charge	Mass /u	Ionising power
alpha	helium nucleus	^4_2He or $^4_2\alpha$	$+2e$	4	high
beta	high-energy electron	$^{0}_{-1}\beta$	$-1e$	$\frac{1}{1840}$	medium
gamma	EM photon (very short λ)	$^0_0\gamma$	neutral	0	low

All of the ionising radiations listed above are emitted from the nucleus of an atom when it undergoes radioactive decay.

Effect of electric and magnetic fields

deflection by an electric field　　deflection by a magnetic field into page

Charged particles are deflected by electric fields. Moving charged particles are deflected by magnetic fields. Uncharged particles are not affected by electric or magnetic fields.

Worked example

Suggest reasons why the ionising power of alpha radiation is much greater than that of beta radiation. **(3 marks)**

Alpha particles are helium nuclei whereas beta particles are electrons. The charge on an alpha particle is double the charge on a beta particle and the alpha particle has much more mass and momentum than a beta particle so it will interact more strongly with molecules in the medium through which it passes, creating more ions per millimetre of its path.

Now try this

1　Suggest a reason why sources of alpha radiation present a low risk to human beings when they are outside the body but are very dangerous if they enter the body. **(3 marks)**

2　A radioactive mineral emits beta and gamma radiation. A student would like to investigate the properties of the gamma radiation alone. How could he do this experimentally? **(2 marks)**

3　Suggest a reason why alpha particles have such a short range in air compared with beta or gamma radiation. **(3 marks)**

Investigating the absorption of gamma radiation by lead

Gamma rays are the most penetrating form of radiation but their intensity can be reduced using a suitable thickness of a dense material.

Gamma radiation in air

Gamma rays are very weakly ionising, so they are barely affected by air, and for distances of a few metres from the source we can assume that they are not absorbed at all. However, they do spread out in all directions from the source, so the intensity of radiation will decline with distance from the source as an inverse square law. This means that in any experiment to measure the absorption of gamma rays the distance between source and detector must be kept constant.

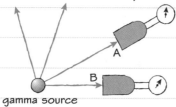

lower counts per minute at A than at
B because of inverse square law

gamma source

To minimise the dose that you receive when using a gamma ray source you must maximise your distance from the source.

Data processing for the experiment

As the thickness of the lead is increased the number of counts per minute C will decrease. If a graph of $\ln C$ is plotted against x it should be a straight line with a negative gradient.

This graph shows that the intensity of the radiation decays exponentially with thickness.

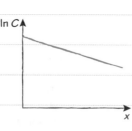

ln C

x

Maths skills If $\ln y$ against x is a straight line with negative gradient, then we know that $\ln y = -mx + c$, so $e^{\ln y} = e^{-m} \times e^c$. The first term is y and the last term is a constant, so $y = Ae^{-bx}$ where A and b are constants.

Investigating the absorption of gamma radiation by lead

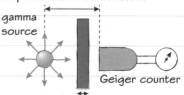

keep this distance constant

gamma source

Geiger counter

lead of thickness x – measure this with a micrometer screw gauge

For each value of lead thickness record the number of counts in a fixed time, e.g. 5 minutes. Repeat and take an average of at least three readings to find the average total counts per minute.

The **independent variable** is the thickness of the lead absorber, x. The **dependent variable** is the number of counts per minute with the source present.

Practical skills Don't forget to measure the average background count with no source present. This must be subtracted from all of the experimental count rates with the source present.

Worked example

Explain why it is important to keep the distance between the source and the detector constant during the experiment described above. **(3 marks)**

Gamma rays from the source are emitted randomly in all directions so they spread out and thus their intensity reduces with distance. If the distance between the source and the detector changed during the experiment, the total count rate would be affected so the experiment would not be a fair test. In order for it to be fair only the independent variable, thickness of lead, must affect the total count rate.

Now try this

1 Suggest a reason why lead is a better absorber of gamma rays than steel or concrete. **(2 marks)**

2 Explain why it is important that the half-life of the gamma ray source used in the experiment above is much longer than the duration of the experiment. **(2 marks)**

Nuclear transformation equations

Nuclear transformation equations must be balanced in order to satisfy three conservation laws.

Three conservation laws

 Conservation of **baryon number**

The number of baryons before and after a nuclear transformation must be the same. Nucleons (protons and neutrons) have baryon number +1. Leptons (electrons and neutrinos) have baryon number zero.

 Conservation of **charge**

The sum of charges on all particles before and after a nuclear transformation must be the same.

 Conservation of **lepton number**

The lepton numbers before and after a nuclear transformation must be the same. Electrons and neutrinos have lepton number +1 and their antiparticles have lepton number −1.

Alpha decay

$$^{238}_{92}U \rightarrow {}^{234}_{90}Th + {}^{4}_{2}He$$

The top line shows the nucleon or baryon number:

✓ 238 = 234 + 4 conserved

The bottom line shows the proton number, or charge on each particle:

✓ 92 = 90 + 2 conserved

There are no leptons on either side of the equation:

✓ 0 = 0 conserved

Gamma decay

Gamma decay does not change the structure of the nucleus; it just removes energy. A gamma ray is a photon with baryon number zero and lepton number zero so it can be emitted in isolation.

Beta decay

$$^{14}_{6}C \rightarrow {}^{14}_{7}N + {}^{0}_{-1}\beta + {}^{0}_{0}\overline{\upsilon}$$

The top line shows the baryon number:

✓ 14 = 14 + 0 + 0 conserved

The bottom line shows the charge on each particle:

✓ 6 = 7 − 1 conserved

There are no leptons on the left, but there is a lepton and anti-lepton on the right.

✓ 0 = 0 + 1 − 1 conserved

Particle creation

$$^{0}_{0}\gamma \rightarrow {}^{0}_{-1}e + {}^{0}_{+1}e$$

✓ Baryon conservation: 0 = 0 + 0

✓ Charge conservation: 0 = −1 + 1

✓ Lepton conservation: 0 = +1 −1

As described on page 88, a photon can produce an electron–positron pair.

Worked example

Copy and complete the transformation equations below:

(a) $^{244}_{94}Pu \rightarrow {}^{4}_{2}He + $ _____   **(2 marks)**

$^{244}_{94}Pu \rightarrow {}^{4}_{2}He + {}^{240}_{92}U$

(b) $^{32}_{15}P \rightarrow$ _____ $+ {}^{0}_{-1}e$ **(2 marks)**

$^{32}_{15}P \rightarrow {}^{32}_{16}S + {}^{0}_{-1}e$

There is a periodic table of elements at the back of this book.

Now try this

1 Radon (Rn) is formed when radium-226 decays by the emission of an alpha particle. Copy and complete the nuclear equation below by adding labels on the right-hand side:

$$^{226}_{88}Ra \rightarrow Rn + \alpha$$ **(4 marks)**

2 Look at the equation for the beta-decay of carbon-14 above.
   (a) Explain why the emitted electron must be accompanied by an antineutrino.   **(3 marks)**
   (b) The carbon-14 nucleus does not contain any electrons. The emitted beta-particle, an electron, is created when a neutron inside the carbon-14 nucleus decays to form a proton. Write down a balanced nuclear transformation equation for the decay of the neutron and show how it satisfies the conservation laws: baryon number, charge and lepton number.   **(5 marks)**

Radioactive decay and half-life

Radioactive decay is a random process governed by chance, but it is still possible to make predictions about the behaviour of large numbers of unstable particles.

Unstable nuclei

If you toss a coin it has a 50% chance of coming down heads. No matter how often you toss the coin or what the previous result was, you always have a 50% chance of getting heads. Radioactive nuclei behave in a similar way. The probability of a particular nucleus decaying in the next minute is the same however long the nucleus has been in existence.

Random: We cannot predict when a particular nucleus will decay. We can only give the probability of decay per unit time, which remains constant until it does decay.

Spontaneous: No external processes, e.g. temperature or pressure, influence the decay of a radioactive nucleus.

Half-life

The **half-life** of a radioactive isotope can be defined in two ways:

1️⃣ The half-life of a radioactive isotope is the time taken for half of the unstable nuclei in a sample of the isotope to decay.

2️⃣ The half-life of a radioactive isotope is the time taken for the **activity** (number of decays per second in the isotope) of a sample to halve.

These are equivalent. If you have twice the number of particles in the source then the activity will double as well:

activity ∝ number of particles in the source

$A \propto N$

The radioactive decay curve

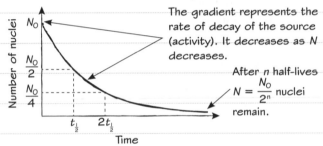

The gradient represents the rate of decay of the source (activity). It decreases as N decreases.

After n half-lives $N = \dfrac{N_0}{2^n}$ nuclei remain.

The graph of activity against time has the same shape and the same half-life.

Radon gas has a half-life of about 1 minute. A sample of radon gas is placed into a sealed container and a detector in the container records an activity of 160 counts per second, after allowing for background radiation.

(a) State what fraction of the original isotope remains after: (i) 1 minute, (ii) 2 minutes, (iii) 5 minutes. **(3 marks)**

(i) 1 minute = 1 half-life so half of the original isotope remains

(ii) 2 minutes = 2 half-lives so $\left(\frac{1}{2}\right)^2 = \frac{1}{4}$

(iii) 5 minutes = 5 half-lives so $\left(\frac{1}{2}\right)^5 = \frac{1}{32}$

(b) Predict the activity after 10 minutes. **(2 marks)**

After 10 half-lives the fraction remaining would be $\left(\frac{1}{2}\right)^{10} \approx 10^{-3}$, so the activity would be about $160 \times 10^{-3} \approx 0$. It is very likely that all of the radon has decayed.

Allowing for background count

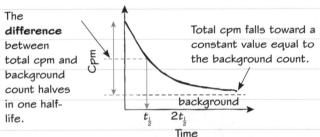

The **difference** between total cpm and background count halves in one half-life.

Total cpm falls toward a constant value equal to the background count.

If you do not subtract the background count from recorded count before plotting, the graph will level off at a constant value.

Carbon-14 is a radioactive isotope of carbon. Carbon-14 has a half-life of 5700 years. All living things contain the same proportion of carbon-14 per gram of carbon in their bodies because it is continuously replaced from the environment. However, when they die, the amount of carbon-14 relative to carbon-12, the common stable isotope of carbon, starts to decrease. This means we can date a sample of dead matter by seeing how much carbon-14 remains.

(a) A piece of an ancient shell necklace contains $\frac{1}{4}$ as much carbon-14 per gram of carbon as a living shellfish. How old is it? **(3 marks)**

(b) A piece of bone found in a cave contains $\frac{1}{16}$ as much carbon-14 as a living thing. How old is the bone? **(2 marks)**

(c) Why can't radiocarbon dating be used to date rocks? **(1 mark)**

Exponential decay

The decay of a radioactive source can be modelled using the exponential function.

Activity

The **activity** of a radioactive source is the number of decays per second taking place in the source:

$$A = \frac{dN}{dt}$$

where A = activity and N = number of nuclei. Conventionally we disregard the fact that the change in N is negative.

S.I. unit: becquerel (Bq)

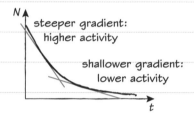

steeper gradient: higher activity

shallower gradient: lower activity

> Activity is the negative of the gradient of the graph of N against time.

Activity is directly proportional to the number of nuclei present, $A \propto N$.

$$A = -\lambda N$$

where λ is the **decay constant**, the probability that a given nucleus will decay per second. The unit is s^{-1}. The negative sign here takes the decline in numbers of nuclei into account.

Exponential decay

Combining the two equations $A = \frac{dN}{dt}$ and $A = -\lambda N$ gives:

$$\frac{dN}{dt} = -\lambda N$$

The solution to this differential equation is:

$$N = N_0 e^{-\lambda t}$$

where N_0 = initial number of nuclei, t = time that has passed and N = number of nuclei remaining at time t. The exponential term is the fraction remaining at time t.

The radioactive decay equation

Using $A = -\lambda N$ we can derive a similar equation for the activity:

$$A = A_0 e^{-\lambda t}$$

where A is activity at time t and A_0 is activity at $t = 0$.

You will find similar exponential equations for charging and discharging capacitors on page 72.

Now try this

The data below show how the activity (corrected for background count) of a source of polonium-218 changes with time over a period of 500 s. By plotting a suitable graph, estimate the half-life of polonium-218. **(4 marks)**

Time/s	Activity/10^{13} s^{-1}
0	1.64
100	1.13
200	0.775
300	0.533
400	0.366
500	0.252

Radioactive decay calculations

Logarithmic plots for the decay of a radioactive source make it easy to find the decay constant.

Half-life

After $t_{\frac{1}{2}}$ the fraction of nuclei remaining is $\frac{1}{2}$, so

$$N = N_0 e^{-\lambda t}$$
$$0.5 = 1 \times e^{-\lambda t_{\frac{1}{2}}}$$
$$e^{-\lambda t_{\frac{1}{2}}} = 0.5$$

Taking natural logarithms of each side and rearranging:

$$\ln e^{-\lambda t_{\frac{1}{2}}} = \ln 0.5$$
$$-\lambda t_{\frac{1}{2}} = \ln 0.5$$
$$\lambda t_{\frac{1}{2}} = -\ln 0.5 = \ln 2$$
$$\lambda = \frac{\ln 2}{t_{\frac{1}{2}}}$$

or $t_{\frac{1}{2}} = \dfrac{\ln 2}{\lambda}$

> **Maths skills** **Natural logarithms and exponentials**
>
> A natural logarithm is a power of e.
> If $y = e^x$ then $\ln y = x$ and $e^{\ln y} = y$
> The usual rules for logarithms apply:
> $$\ln(a \times b) = \ln(a) + \ln(b)$$
> $$\ln\left(\frac{a}{b}\right) = \ln(a) - \ln(b)$$
> These relations are very useful when rearranging equations for exponential decay, e.g. radioactivity, capacitors.

Analysing radioactive decay

In many experiments you will have values of N or A and t and need to find λ and $t_{\frac{1}{2}}$. The best way to do this is to use natural logarithms.

$$N = N_0 e^{-\lambda t}$$
$$\ln N = \ln N_0 - \lambda t$$
$$y = c + mx$$

1 year = 365.25 days

graph: $\ln N$ vs t, gradient $= -\lambda$, intercept $\ln N_0$

Worked example

A radioactive isotope X has a half-life of 2.0 years.
(a) Calculate its decay constant in s^{-1}.   **(2 marks)**

$$\lambda = \frac{\ln 2}{t_{\frac{1}{2}}} = \frac{\ln 2}{2.0 \times 365.25 \times 24 \times 3600}$$
$$= 1.1 \times 10^{-8}\,s^{-1}$$

(b) A rock contains 2.00×10^{18} atoms of X. Calculate the activity of the rock.   **(2 marks)**

$$A = -\lambda N = 1.1 \times 10^{-8} \times 2.00 \times 10^{18}$$
$$= 2.2 \times 10^{10}\,Bq$$

> The minus sign in this formula means the number of nuclei in the sample, N, decreases with time. In practice we can ignore the sign when using the formula.

(c) Calculate how many atoms of X will remain in the rock above after:
   (i) 10 years   **(1 mark)**
   (ii) 1 year   **(2 marks)**

(i) 10 years is 5 half-lives so $N = \dfrac{2.00 \times 10^{18}}{2^5}$
$$= 6.3 \times 10^{16}$$

(ii) $N = N_0 e^{-\lambda t}$
$$= 2.00 \times 10^{18} \times e^{-(1.1 \times 10^{-8} \times 1.0 \times 365.25 \times 24 \times 3600)}$$
$$= 1.4 \times 10^{18} \text{ atoms of X}$$

Now try this

1 The data below, already used on the previous page, show how the activity, corrected for background count, of a source of polonium-218 changes with time over a period of 500 s. By plotting a suitable logarithmic graph find the decay constant and half-life of polonium-218   **(6 marks)**

Time/s	Activity/10^{13} s^{-1}
0	1.64
100	1.13
200	0.775
300	0.533
400	0.366
500	0.252

2 Americium-241, used in domestic smoke alarms, has a decay constant $\lambda = 5.1 \times 10^{-11}\,s^{-1}$. Calculate its half-life in years, and how long it would take for its activity to drop to 5% of the original activity.   **(4 marks)**

Gravitational fields

Gravitational forces attract masses to one another, because each mass creates a field in space that affects other masses.

Fields and forces

Two separate masses exert attractive forces on one another. These forces depend on the masses and the distance between them.

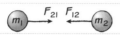

$F_{12} = F_{21}$ by Newton's third law.

The force on each mass is explained by assuming that each mass creates a field in the space around it, and that this field exerts a force on the other mass.

A **gravitational field** is defined as a region where a mass experiences a force.

Gravitational field strength, g

The gravitational force on a mass is called its **weight**. Weight depends on the mass and the strength of the gravitational field at the position of the mass.

$$W = mg$$

where W = weight in newtons (N), m = mass in kilograms (kg) and g = gravitational field strength (N kg^{-1}).

Gravitational field strength g is the gravitational force per unit mass at a point in the field.

g = gravitational force/mass (S.I. unit: N kg^{-1})

$$g = \frac{F}{m}$$

Near the surface of the Earth, $g = 9.81$ N kg^{-1}

Near the surface of the Moon, $g = 1.63$ N kg^{-1}

Gravitational field lines

The field lines (arrows) represent the direction of gravitational force exerted on a mass placed in the field. The field is strongest where the field lines are closest together.

Uniform gravitational fields

The field near the surface of a planet or star is approximately uniform.

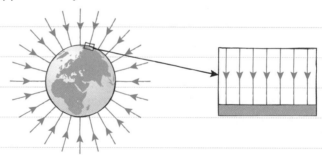

A uniform field is represented by parallel, equally spaced field lines; g = constant.

Worked example

Calculate the weight of a 75 kg man on the surface of the Earth and on the surface of the Moon. **(2 marks)**

$W = mg = 75 \times 9.81 = 740$ N on Earth.

$W = mg = 75 \times 1.63 = 120$ N on the Moon.

Now try this

1 Calculate the weight of a 60 kg woman on the surface of the Earth and on the surface of the Moon. **(2 marks)**

2 Explain why the gravitational field near the surface of the Earth is treated as uniform in calculations even though the field around the planet is radial. **(2 marks)**

Gravitational potential and gravitational potential energy

As an object falls towards the ground, its gravitational potential decreases.

Energy in the gravitational field

When a mass moves through a gravitational field, work must be done by or against gravitational forces. This changes the gravitational potential energy of the mass.

Lifting a mass in a uniform gravitational field

To lift a mass m from A to B a force $F = mg$ must move through a distance h.

Work done $= mg\Delta h$

$mg\Delta h =$ increase in gravitational potential energy
$= \Delta GPE$

Gravitational potential

Gravitational potential $V_g =$ GPE per unit mass

S.I. unit: $J\,kg^{-1}$

When a mass is lifted through a height h in a uniform field of strength g:

$$\Delta GPE = mg\Delta h$$
$$\Delta V_g = \frac{mg\Delta h}{m} = g\Delta h$$

The zero of gravitational potential is defined to be at infinity, though, for convenience, we often redefine the zero for GPE to suit the circumstance – just like electrical potential. If a mass was lifted from the Earth's surface out to such a distance that it experienced no gravitational attraction, effectively infinite distance from Earth, then its GPE would have increased to zero. This means that all absolute gravitational potentials for masses in the Universe must be negative.

Field lines and equipotentials

Equipotential surfaces are perpendicular to gravitational field lines. No work has to be done by or against gravitational forces to move a mass along an equipotential.

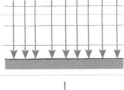

Worked example

(a) Calculate the increase in gravitational potential energy when a 75 kg man climbs a vertical distance of 12 m (i) on Earth and (ii) on the Moon. **(2 marks)**

(i) $\Delta GPE = mg\Delta h = 75 \times 9.81 \times 12$
$= 8800\,J$

(ii) $\Delta GPE = mg\Delta h = 75 \times 1.63 \times 12$
$= 1500\,J$

(b) Calculate the increase in gravitational potential for questions (a) (i) and (ii). **(2 marks)**

(i) $\Delta V_g = g\Delta h = 120\,J\,kg^{-1}$ on Earth

(ii) $\Delta V_g = g\Delta h = 20\,J\,kg^{-1}$ on the Moon

Now try this

1 (a) Calculate the increase in gravitational potential energy when a 60 kg woman climbs a vertical distance of 25 m (i) on Earth, and (ii) on the Moon. **(2 marks)**
   (b) Calculate the increase in gravitational potential for questions (a) (i) and (ii). **(2 marks)**

2 Explain why the equipotentials in a uniform field are equally spaced. **(3 marks)**

Newton's law of gravitation

Newton's inverse square law describes the attractive force between all masses in the Universe.

Newton's law of gravitation

Two point masses m_1 and m_2 separated by a distance r attract one another with a force F that is directly proportional to the product of the masses and inversely proportional to the square of their separation.

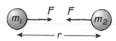

$$F \propto m_1 m_2$$

$$F \propto \frac{1}{r^2}$$

These can be combined to give **Newton's law of gravitation**:

$$F = \frac{Gm_1 m_2}{r^2}$$

where G is the universal gravitational constant

$G = 6.67 \times 10^{-11}\,\mathrm{N\,m^2\,kg^{-2}}$

Point masses

Spherical objects such as planets and stars can be treated as if all of their mass is concentrated at their centre of mass, like a point.

If you are calculating the gravitational force on a non-spherical object, e.g. a human, it is often possible to treat that as a point mass too!

Combining gravitational forces

If a mass is affected by the gravitational fields of more than one other mass the resultant force is found by vector addition of the forces.

Newton's third law and gravitation

The gravitational forces on a pair of objects form an action–reaction pair and are equal and opposite. For example, for a person standing on the surface of the Earth:

gravitational force of Earth on person
= (−)gravitational force of person on Earth.

In other words, the person's gravitational field attracts the Earth as strongly as the Earth's gravitational field attracts the person.

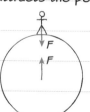

Don't forget to convert km to m.

Worked example

1 Calculate the gravitational force exerted by the Moon on the Earth. (mass of the Moon $= 7.3 \times 10^{22}\,\mathrm{kg}$, mass of the Earth $= 6.0 \times 10^{24}\,\mathrm{kg}$, radius of the Moon's orbit $= 380\,000\,\mathrm{km}$, $G = 6.67 \times 10^{-11}\,\mathrm{N\,m^2\,kg^{-2}}$) **(2 marks)**

$$F = \frac{Gm_1 m_2}{r^2}$$

$$= 6.67 \times 10^{-11} \times 7.3 \times 10^{22} \times \frac{6.0 \times 10^{24}}{(3.8 \times 10^8)^2}$$

$$= 2.0 \times 10^{20}\,\mathrm{N}$$

2 State the value of the gravitational force exerted by the Earth on the Moon. **(1 mark)**

$2.0 \times 10^{20}\,\mathrm{N}$, by Newton's third law

Now try this

(a) Use Newton's law of gravitation to calculate the weight of a 60 kg woman standing on the surface of the Earth. (radius of the Earth $r_E = 6400\,\mathrm{km}$, mass of the Earth $= 6.0 \times 10^{24}\,\mathrm{kg}$, $G = 6.67 \times 10^{-11}\,\mathrm{N\,m^2\,kg^{-2}}$) **(3 marks)**

(b) Define the gravitational field strength g at the surface of the Earth. **(2 marks)**

(c) Use your answer to (a) to calculate a value for the gravitational field strength at the surface of the Earth. **(2 marks)**

Gravitational field of a point mass

Point masses create radial gravitational fields.

Field strength due to a point mass

If a small mass δm is placed in the field of a point mass m at a distance r from m there will be a gravitational force F:

$$F = \frac{Gm\delta m}{r^2}$$

Gravitational field strength g is defined as force per unit mass at a point in the field. Therefore, the gravitational field strength at a point in space at distance r from a point mass m is:

$$g = \frac{Gm\delta m}{\delta m r^2}$$

$$g = \frac{Gm}{r^2} \text{ (units: N kg}^{-1})$$

The field strength obeys an inverse square law just like the force.

Radial fields

The gravitational field around a point mass is radial – all the field lines point inwards toward the central point. The field due to a uniform spherical mass is radial, too, but the equation for field strength can only be used outside the surface.

For a spherical mass,

$$g = \frac{Gm}{r^2} \text{ for } r > R$$

The field strength at the surface is given by:

$$g = \frac{Gm}{R^2}$$

Variation of gravitational field strength due to the Earth

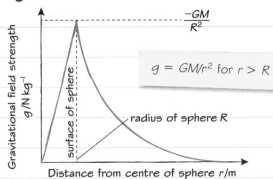

The equipotentials get closer together as the field gets stronger closer to the Earth.

The graph above shows the theoretical variation of g from the centre of the Earth out to large distances. Beyond the surface g obeys an inverse square law. This graph treats the Earth as a sphere of uniform density; in reality the density is greater close to the core.

Worked example

Calculate the gravitational field strength at the altitude of the Hubble Space telescope, 560 km. (The radius of the Earth is 6400 km and its mass is 6.0×10^{24} kg.) **(2 marks)**

$$g = \frac{Gm}{r^2} = 6.67 \times 10^{-11} \times 6.0 \times \frac{10^{24}}{(6\,960\,000)^2}$$

$$= 8.3 \text{ N kg}^{-1}$$

Altitude h is distance above the surface of the Earth. The distance from the centre of the Earth is altitude plus the radius of the Earth, $h + R_E$.

Now try this

1 Calculate the gravitational field strength due to the Earth's field at the radius of the Moon's orbit, 380 000 km. **(2 marks)**

2 The gravitational field strength at the surface of the Moon is about 1.6 N kg^{-1}. What is the strength of the Moon's field at distances R and $4R$ from the surface (where R is the radius of the Moon)? **(4 marks)**

3 Explain why the gravitational field strength at the centre of the Earth is zero. **(2 marks)**

Gravitational potential in a radial field

Radial fields are particularly important because most astronomical bodies, such as the Earth, are spherical and thus have radial gravitational fields.

Gravitational potential V_{grav}

Refer back to page 120 for reminders that:

- All gravitational potentials are measured relative to the potential at infinity, which is taken to be zero. This means that if all masses in the Universe were completely separated from one another then they would have zero potential energy.

- Gravity is universally attractive, so work must be done to move a mass away to infinity. This means that gravitational potential energy increases as masses are separated. Therefore, all gravitational potential energies must be negative.

- The **gravitational potential** V_{grav} at a point in a gravitational field is the work that must be done per unit mass (per kg) to move a small mass from infinity and place it at that point in the field.

$$V_{grav} = GPE/m \text{ (S.I. unit: } J\,kg^{-1})$$

Gravitational potential in a radial field

The gravitational potential at a point a distance r from a point mass M is given by:

$$V_{grav} = \frac{-GM}{r}$$

For a uniform spherical mass of radius R this formula only applies for points outside the surface of the mass ($r > R$).

> **Be careful!** The formula for gravitational potential looks very similar to the formula for gravitational field strength. However:
> $$V_{grav} \propto \frac{1}{r}$$
> $$g \propto \frac{1}{r^2}$$

Calculating the change in GPE near a spherical mass

If a mass m is moved from a distance r_1 to a greater distance r_2 away from a point or spherical mass M the work done is calculated using:

$$W = \Delta GPE = m\Delta V_{grav}$$
$$= m\left(\frac{-GM}{r_2}\right) - \left(\frac{-GM}{r_1}\right) = GMm\left(\frac{1}{r_1} - \frac{1}{r_2}\right)$$

Gravitational potential energy

$$V_{grav} = \frac{GPE}{m}$$

so the gravitational potential energy of a mass m placed at a point r from a mass M is:

$$GPE = V_{grav}m = \frac{-GMm}{r}$$

> Be careful when calculating changes in GPE. The formula $\Delta GPE = mgh$ is only valid when g can be assumed constant, e.g. for short vertical distances near the surface of the Earth.
>
> In general $\Delta GPE = m\Delta V_{grav}$.

Calculate the gravitational potential at the Earth's surface. (radius of the Earth = 6400 km, mass of the Earth = 6.0×10^{24} kg)   **(3 marks)**

$$V_{grav} = \frac{-GM}{r_E}$$
$$= -6.67 \times 10^{-11} \times 6.0 \times \frac{10^{24}}{6.4 \times 10^6}$$
$$= -6.3 \times 10^7 \, J\,kg^{-1}$$

Gravitational potential near the Earth's surface

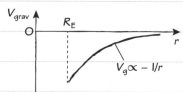

Now try this

(a) Calculate the gravitational potential at the altitude of the Hubble Space telescope, 590 km above the Earth's surface. (radius of the Earth = 6400 km, mass of the Earth = 6.0×10^{24} kg)   **(2 marks)**

(b) Calculate the work that would need to be done to lift the space telescope, of mass about 10 000 kg, from the surface of the Earth to an altitude of 590 km.   **(4 marks)**

(c) Explain why, in fact, a great deal more energy than this was required to put the space telescope into orbit.   **(3 marks)**

Energy changes in a gravitational field

When a mass moves in a gravitational field energy can be transferred between the mass and the field.

Moving in radial and uniform fields

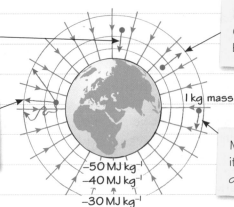

Mass moves inwards: its GPE decreases and work is done by the gravitational field.

Mass is moved outwards: its GPE increases and work must be done by an external agent.

ΔGPE = + 20 MJ regardless of the route it takes between the two equipotentials.

1 kg mass

Mass moves along an equipotential: its GPE is constant and no work is done.

−50 MJ kg⁻¹
−40 MJ kg⁻¹
−30 MJ kg⁻¹

GPE is mapped by equipotential lines. Note that the equipotentials get progressively farther apart. This is because the field strength gets weaker with distance from the Earth and **gravitational potential** $V_{grav} \propto 1/r$

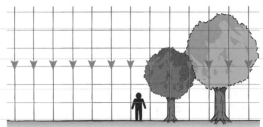

Maximum gravitational potential energy, minimum kinetic energy.

If a mass is thrown upwards, as it decelerates its KE reduces and its GPE increases. As it falls back down the GPE decreases and KE increases. Note that the equipotentials are equally spaced because the field strength is approximately constant. This is because the height that the ball reaches is tiny compared with the radius of the Earth.

Worked example

A 50 kg mass is moved from a position with gravitational potential −52 MJ kg⁻¹ to a position with gravitational potential −48 MJ kg⁻¹.

(a) Calculate the change in gravitational potential energy. **(2 marks)**

$\Delta GPE = m\Delta V_{grav} = 50(-48 - -52)$
$= 50(+4) = 200\,MJ$

(b) Explain whether the gravitational field does work or not. **(2 marks)**

GPE has increased as the mass has been pushed upwards, so work has been done by an external agent and not by the field.

The potential values were left in MJ kg⁻¹, so the answer is in MJ, not J.

Now try this

The gravitational potential at the Earth's surface is about −57 MJ kg⁻¹.

(a) How much work must be done to lift a mass of 1.0 kg from the surface of the Earth to a great distance away from the Earth (effectively to infinity)? **(2 marks)**

(b) Imagine that a mass of 1.0 kg were to be given enough kinetic energy to just escape from the surface of the Earth. What is the minimum velocity required for it to escape? **(4 marks)**

(c) Explain why the 'escape velocity' calculated for a 1.0 kg mass in (b) would be the same for any mass. **(2 marks)**

Comparing electric and gravitational fields

Electric and gravitational fields have a lot in common, but there are some important differences.

Comparing key aspects of electric and gravitational fields

	Gravitational field	Electric field
range of field	infinite	infinite
source of field	mass	charge
objects affected by field	massive particles	charged particles
field strength	$g = F/m$ (N kg^{-1})	$E = F/q$ (N C^{-1} or V m^{-1})
potential	$V_{grav} = GPE/m$ (J kg^{-1})	$V = EPE/q$ (J C^{-1})
work done in field	$W = m\Delta V_{grav}$ (J)	$W = q\Delta V$ (J)
force exerted	always attractive	like charges repel, unlike charges attract
law of force between particles	inverse square law	inverse square law
potential energy	always negative	can be positive or negative

GPE = gravitational potential energy; EPE = electrostatic potential energy

Key difference

There are two different types of charge, positive and negative, but only one type of mass.

This makes all gravitational forces attractive, while electrostatic forces can be attractive or repulsive.

This also means that gravitational effects increase with mass whereas electrostatic effects tend to cancel out because matter is, on average, neutral.

Comparing radial fields

	Gravitational	Electrostatic
force law	$F = Gm_1m_2/r^2$	$F = \dfrac{Q_1Q_2}{4\pi\varepsilon_0 r^2}$
field strength	$g = Gm/r^2$	$E = \dfrac{Q}{4\pi\varepsilon_0 r^2}$
potential	$V_{grav} = -Gm/r$	$V = \dfrac{Q}{4\pi\varepsilon_0 r}$
potential energy	$GPE = -Gm_1m_2/r$	$EPE = \dfrac{Q_1Q_2}{4\pi\varepsilon_0 r}$
potential at infinity	zero	zero

The field and equipotential patterns for a mass and a negative charge are similar. In both cases field lines point toward the centre and equipotentials increase as they get further from the centre. The opposite is true for a positive charge.

Worked example

When a mass is placed a finite distance from another mass its gravitational potential energy is always negative. Under what circumstances can the electrostatic potential energy of a charge be positive?
(2 marks)

If a positive charge is placed a finite distance from another positive charge its electrostatic potential energy is positive and if a negative charge is placed a finite distance from another negative charge its electrostatic potential energy is positive.

Now try this

1 The gravitational attraction between the proton and the electron in a hydrogen atom is about 10^{39} times weaker than the electrostatic attraction between the same two particles. How would this ratio change if the particles were ten times farther apart? Explain your answer. **(3 marks)**

2 All masses have the same acceleration in a uniform gravitational field but not all charges have the same acceleration in a uniform electric field. Explain this difference. **(4 marks)**

Orbits

To understand orbital motion we need to use Newton's laws of motion and gravitation together.

A circular orbit

Newton's first law: In the absence of a resultant force a body continues to move at constant velocity in a straight line.

When a body moves in a circular orbit its velocity vector is continually changing direction. This is a **centripetal acceleration** and is caused by a **centripetal force**.

The centripetal force for orbital motion is provided by the gravitational attraction between the orbiting body and the body at the centre of its orbit (e.g. between the Moon and the Earth).

Types of Earth orbit

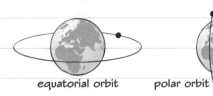

equatorial orbit   polar orbit

> Satellites can orbit the Earth in any plane, at any distance and in any direction, but their orbital period is determined by the distance.

Geostationary satellites are placed in equatorial orbits at a distance such that their period of orbit is 24 hours. This keeps them stationary relative to a point on the Earth's equator, so they are ideal for use as communications satellites.

Gravity and centripetal force for a circular orbit

Gravitational attraction = centripetal force

$$F_{grav} = \frac{mv^2}{r} = mr\omega^2$$

where r is the radius of the orbit and v is the orbital speed.

Thus $\dfrac{Gm_1m_2}{r^2} = \dfrac{m_2v^2}{r}$

To review centripetal acceleration and force, look at page 65.

The time period of an orbit is T.

$$T = \text{circumference/speed} = \frac{2\pi r}{v}$$

so $v = \dfrac{2\pi r}{T}$

Substitute this into the force equation and rearrange to get:

$$\frac{r^3}{T^2} = \frac{Gm_1}{4\pi^2}$$

This ratio $\left(\dfrac{r^3}{T^2}\right)$ will be a constant value for all satellites of the same central body, for example all planets orbiting the Sun.

Elliptical orbits

Planets do not orbit in perfect circles. The shape of a planetary orbit is an **ellipse**. Comets too have elliptical orbits but these are much more elongated, eccentric, than planetary orbits.

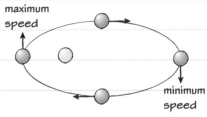

maximum speed

minimum speed

The gravitational potential energy increases as the planet or comet moves farther from the Sun. However, its total energy is constant, so its kinetic energy falls and thus its speed varies around the orbit.

Worked example

Show that the time period T for a satellite orbit increases with its radius r. **(3 marks)**

Let the mass of the Earth be m_1, the mass of the satellite be m_2 and the radius of the orbit be r.

$$\frac{Gm_1m_2}{r^2} = \frac{m_2v^2}{r}$$

m_2 cancels and $v = \dfrac{2\pi r}{T}$, so

$$\frac{Gm_1}{r^2} = \frac{4\pi^2r^2}{rT^2}$$

$$T^2 = \frac{4\pi^2r^3}{Gm_1}$$

$T^2 \propto r^3$, so as r increases so does T.

Now try this

1 Calculate the radius of orbit of a geostationary satellite, i.e. a satellite with time period 24 hours. (mass of Earth = 6.0×10^{24} kg; $G = 6.66 \times 10^{-11}$ N m^2 kg^{-2}) **(4 marks)**

2 Explain why a satellite cannot be placed into a geostationary polar orbit. **(2 marks)**

Exam skills 12

This exam-style question uses knowledge and skills you have already revised. Have a look at pages 119 and 126, for a reminder about orbits and gravitational fields.

Worked example

The Hubble Space Telescope orbits Earth in a circular orbit at an altitude of 590 km. The mass of the Hubble Space Telescope is 11 000 kg. The mass of the Earth is 6.0×10^{24} kg and the radius of the Earth is 6400 km. ($G = 6.67 \times 10^{-11}$ N m^2 kg^{-2})

(a) Suggest and explain one advantage of putting a telescope into orbit around the Earth. **(2 marks)**

The telescope can obtain clearer and more detailed images of astronomical objects because the light it receives does not have to pass through the Earth's atmosphere, which will absorb some photons.

(b) Calculate the gravitational potential energy of the Hubble Space Telescope. **(2 marks)**

$$GPE = \frac{-GMm}{r}$$

$$= \frac{-6.67 \times 10^{-11} \times 6.0 \times 10^{24} \times 11\,000}{6.40 \times 10^6 + 0.59 \times 10^6}$$

$$= -6.3 \times 10^{11} \text{ J}$$

(c) Calculate the time period of the orbit of the Hubble Space Telescope. **(3 marks)**

$$\frac{GMm}{r^2} = mr\omega^2 = \frac{4\pi^2 mr}{T^2}$$

$$T = R\left(\frac{4\pi^2 r^3}{GM}\right) = 5804 \text{ s} = 97 \text{ minutes.}$$

(d) (i) Calculate the total energy of the Hubble Space Telescope in its orbit. **(4 marks)**

Total energy =
gravitational potential energy + kinetic energy

$$E_K = \tfrac{1}{2}mv^2 = \tfrac{1}{2}m\left(\frac{2\pi r}{T}\right)^2$$

$$= \tfrac{1}{2} \times 11\,000 \times \left(2\pi \times 6.99 \times \frac{10^6}{5804}\right)^2$$

$$= +3.15 \times 10^{11} \text{ J}$$

Total energy = $-6.3 \times 10^{11} + 3.15 \times 10^{11}$
$= -3.15 \times 10^{11}$ J

(d) (ii) Comment on the sign of the total energy and the relative sizes of the GPE and KE. **(2 marks)**

The total energy is negative because the satellite is trapped in an orbit – it does not have enough energy to escape from the Earth's gravitational field, to reach zero total energy. The kinetic energy is minus one half of the gravitational potential energy.

Be careful – altitude is measured from the surface of the Earth, but the distance needed when using the gravitational field equations must be measured from the centre of the Earth: $r = R_E + h$
$= 6.4 \times 10^6 + 0.59 \times 10^6 = 6.99 \times 10^6$ m

When a question asks you to 'suggest and explain' it is expecting you to bring your knowledge of physics to bear in an unfamiliar context. Make sure that you answer both parts of a question like this – back up your suggestion with a valid explanation.

Look at how this answer has been structured:
 Quote the relevant equation.
2 Substitute values into the equation.
3 Carry out the calculation and give the answer.
4 Include correct S.I. units.
5 Watch out for the sign!

When you carry out a more complex calculation it is helpful to you, and to the examiner, if you show each stage of your working, using algebra to highlight the key steps.

Paying attention to getting the signs right is crucial here!

This question asks you to comment on the sign, so it is sensible to go back at this stage and check that you have not made a sign error or lost a minus sign in the earlier parts. Here 3 s.f. have been used to emphasise the fact that the magnitude of the kinetic energy is exactly half the gravitational potential energy.

Simple harmonic motion

Simple harmonic motion provides a powerful mathematical model of vibration, from atomic oscillations in crystals to the vibrations of stringed instruments.

Oscillations

An oscillation is a regular repeated motion around a fixed point. Its characteristics are:

- time period T – the time for one complete cycle of oscillation (s)
- frequency f – the number of oscillations per second (Hz): $f = 1/T$
- amplitude A – the maximum displacement of the oscillator from its equilibrium position (m).

An example is a mass on a spring:

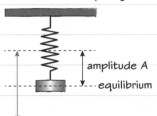

amplitude A

equilibrium

One complete oscillation is from equilibrium to the highest point, back down to the lowest point and back up to equilibrium.

Simple harmonic oscillations

Simple harmonic oscillations come about when two conditions are met:

1 There is a force F on the oscillator which is directed back toward the equilibrium position.

2 The magnitude of the force F is directly proportional to the displacement x from equilibrium.

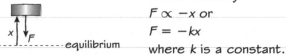

x F ⟶ equilibrium

Mathematically:

$F \propto -x$ or

$F = -kx$

where k is a constant.

Proving an oscillator is simple harmonic

Find the force acting on the oscillator.

Show that it acts toward the equilibrium position and is directly proportional to distance from equilibrium.

🧪 Practical skills — Measuring the time period of an oscillator

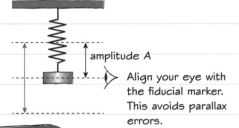

Place a fiducial marker (e.g. an optical pin) level with the equilibrium position.

amplitude A

Align your eye with the fiducial marker. This avoids parallax errors.

- Set the oscillator going and measure the time for n complete oscillations (nT).
- Count each oscillation as the oscillator moves past its equilibrium position in the same direction, e.g. moving downwards. It moves fastest at this position so the exact time at which it passes is easier to judge than at the amplitudes where it moves slowly.
- Repeat this to obtain a minimum of three results. This reduces random errors and allows you to spot any anomalies, e.g. if you miscounted.
- Hence find the average value of nT. Divide by n to find the average time for one oscillation.

Worked example

A trolley is held by an identical horizontal spring at each end, both of which obey Hooke's law, with individual spring constants equal to k. Show that, when the trolley is displaced horizontally a small distance and then released, it will undergo simple harmonic oscillations. **(4 marks)**

When the trolley is displaced a distance x from equilibrium the tension in one spring increases by kx and the tension in the other spring decreases by kx. The resultant force on the trolley is $2kx$ and the direction of the force is toward equilibrium:

$F = -2kx$

Force is therefore directly proportional to displacement and directed toward equilibrium. These are the conditions for simple harmonic motion, so the trolley does undergo simple harmonic oscillation.

Now try this

A student measures the time for 10 complete oscillations of a pendulum and obtains the following results: 11.12 s, 11.22 s, 11.18 s, 10.06 s, 11.15 s

(a) State which of these measurements is anomalous. **(1 mark)**

(b) Suggest and justify a reason for your answer to (a). **(2 marks)**

(c) Use the data to calculate the average time for one oscillation. **(3 marks)**

Analysing simple harmonic motion

Simple harmonic oscillations can be described mathematically with the sine and cosine functions.

Acceleration and angular frequency

For simple harmonic motion to occur the force on the oscillator must be of the form:

$$F = -kx$$

The acceleration of the oscillator is given by:

$$a = -(k/m)x$$

It is convenient to put (k/m) equal to ω^2. This simplifies later equations, when ω can be related to frequency f:

$$a = -\omega^2 x$$

This is called the equation of motion for a simple harmonic oscillator, and:

$$\omega = 2\pi f = \frac{2\pi}{T}$$

ω is called the angular frequency and has units s^{-1}.

The time period and frequency of a simple harmonic oscillator are both independent of its amplitude.

Displacement, velocity and acceleration

The displacement of a simple harmonic oscillator can be described by the equation:

$$x = A \cos \omega t$$

where x = displacement and A = amplitude.

The velocity of the same oscillator is given by:

$$v = -\omega A \sin \omega t$$ — The velocity is 90° out of phase with the displacement.

where v = velocity and ωA = maximum velocity.

The acceleration of the same oscillator is given by:

$$a = -\omega^2 A \cos \omega t$$ — The acceleration is 90° out of phase with the velocity and 180° out of phase with the displacement.

where a = acceleration and $\omega^2 A$ = maximum acceleration.

Kinetic energy of a simple harmonic oscillator

Kinetic energy (KE) $= \frac{1}{2}mv^2$

For a simple harmonic oscillator:

$$v = -\omega A \sin \omega t$$

$$KE = \frac{1}{2}m\omega^2 A^2 \sin^2 \omega t$$

The maximum value of a sine is 1, so the maximum kinetic energy is $\frac{1}{2}m\omega^2 A^2$.

This is also equal to the total energy of the oscillator:

Total energy $= \frac{1}{2}m\omega^2 A^2$

Notice that this is proportional to A^2.
This is the reason why wave intensity is proportional to A^2.

Worked example

A mass of 0.50 kg undergoes simple harmonic motion with amplitude 0.020 m and frequency 1.5 Hz.

(a) Calculate its maximum velocity and state at which position in the oscillation this occurs. **(3 marks)**

$v_{max} = \omega A = 2\pi f A = 2 \times \pi \times 1.5 \times 0.020 = 0.19 \, m\,s^{-1}$. This occurs as the oscillator passes its equilibrium position.

(b) Calculate the maximum acceleration and state at which positions this occurs. **(2 marks)**

$a_{max} = \omega^2 A = 4 \times \pi^2 \times (1.5)^2 \times 0.020 = 1.8 \, m\,s^{-1}$. This occurs when the oscillator is at each maximum amplitude.

(c) What is the maximum force exerted on the oscillator during each oscillation and where does this occur? **(2 marks)**

$F_{max} = ma_{max} = 0.50 \times 1.8 = 0.90 \, N$. This occurs where acceleration is maximum, i.e. when the oscillator is at each maximum amplitude.

Now try this

A trolley of mass 0.60 kg, held by an identical horizontal spring at each end, both of which obey Hooke's law, is displaced horizontally a distance of 2.5 cm and released. The initial resultant force on the trolley is 0.80 N.

(a) Calculate the initial acceleration of the trolley. **(1 mark)**

(b) Calculate the angular frequency of the oscillations. **(1 mark)**

(c) Calculate the frequency of the oscillations. **(1 mark)**

(d) Calculate the maximum speed of the trolley during one oscillation and state where this occurs. **(3 marks)**

Graphs of simple harmonic motion

Representing oscillations graphically makes it much easier to see how displacement, velocity and acceleration are related.

Capturing simple harmonic motion using a position sensor

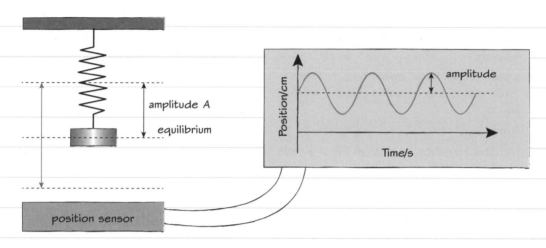

Graphs of motion

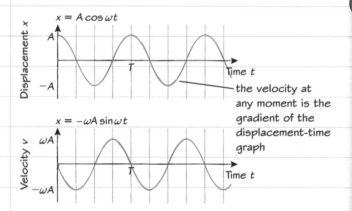

the velocity at any moment is the gradient of the displacement–time graph

The gradient of the displacement–time graph is equal to the velocity of the oscillator at that moment, and the gradient of the velocity–time graph is equal to the acceleration of the oscillator at that moment.

The graph below shows displacement versus time for a simple harmonic oscillator:

(a) What is the amplitude and the frequency of this oscillator? **(3 marks)**

Amplitude = 4.5 cm. Time period = $\dfrac{0.080}{2}$

= 0.040 s, so $f = \dfrac{1}{T} = \dfrac{1}{0.040} = 25\,Hz$

(b) At what times does the oscillator have maximum positive velocity? **(2 marks)**

Maximum positive velocity corresponds to maximum positive gradient, so the times are 0.030 s and 0.070 s.

Sketch a graph to show the acceleration against time for the oscillator in the worked example above. Include a value for the maximum acceleration on the acceleration axis. **(5 marks)**

The mass–spring oscillator and the simple pendulum

The mass–spring oscillator is a typical simple harmonic oscillator and is used as a model for many different types of oscillation in nature.

Frequency and time period for a mass–spring oscillator

A mass–spring oscillator consists of a mass m suspended from a light spring. The spring obeys Hooke's law and has spring constant k, so the force law for the oscillator is:

$$F = -kx$$

Therefore:

 the force is directly proportional to displacement from equilibrium

 the force is always directed back toward equilibrium.

In other words, a mass–spring oscillator is a simple harmonic oscillator.

The equation of motion for all simple harmonic oscillators is:

$$a = -\omega^2 x$$

$$\omega^2 = \frac{ak}{F}$$

$$\omega = \sqrt{\left(\frac{k}{m}\right)}$$

$$f = \frac{1}{2}\pi\sqrt{\frac{k}{m}} \text{ and } T = 2\pi\sqrt{\frac{m}{k}}$$

Frequency and time period for a simple pendulum

A simple pendulum consists of a small mass suspended from a light, inextensible string. When it is displaced sideways it undergoes oscillations.

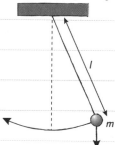

The force accelerating the pendulum bob is due to the tangential component of its weight. For small oscillations this is approximately directly proportional to the displacement and directed back toward equilibrium, so a simple pendulum is a simple harmonic oscillator for **small amplitudes** (less than 10 degrees from the rest position) only.

Its frequency and time period are given by:

$$f = \frac{1}{2\pi}\sqrt{\frac{g}{l}} \text{ and } T = 2\pi\sqrt{\frac{l}{g}}$$

🧪 Practical skills — Finding the time period of a simple pendulum

1 Set up a fiducial marker at the equilibrium position so that the pendulum passes in front of it.

2 Align your eye in a straight line in front of the pendulum and fiducial marker.

3 Displace the pendulum though a small angle, less than 10°, and release it.

4 Time n oscillations, e.g. 20.

5 Repeat several times.

6 Remove any anomalous results and find the average time for n oscillations.

7 Divide by n to find the average time period for one oscillation.

Worked example

1 A mass–spring oscillator has a time period of 0.80 s and a spring constant of 20 N m⁻¹. Calculate the mass of the oscillator. **(2 marks)**

$T = 2\pi\sqrt{\left(\frac{m}{k}\right)}$ so $T^2 = 4\pi^2\frac{m}{k}$, therefore

$m = \frac{kT^2}{4\pi^2} = 0.32\,\text{kg}$

2 Calculate the frequency of a simple pendulum of length 1.0 m. **(2 marks)**

$f = \frac{1}{2\pi}\sqrt{\left(\frac{g}{l}\right)} = \frac{1}{(2\pi)}\sqrt{\left(\frac{9.81}{1.0}\right)} = 0.50\,\text{s}$

Now try this

1 Calculate the length of a simple pendulum that would have the same time period as the mass–spring oscillator in the worked example, 0.80 s. **(3 marks)**

2 A mass–spring oscillator and a simple pendulum both have time periods of 1.0 s when they oscillate on Earth. State and explain qualitatively how their time periods will change (if at all) when they oscillate on the Moon, where g is much lower. **(4 marks)**

Energy and damping in simple harmonic oscillators

Oscillators transfer energy between kinetic and potential, but they usually also dissipate energy to their surroundings.

Energy transfers for an undamped mass spring oscillator

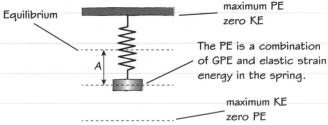

Equilibrium

maximum PE
zero KE

The PE is a combination of GPE and elastic strain energy in the spring.

A

maximum KE
zero PE

Provided the oscillator is undamped – see below – by conservation of energy the quantitative energy transfers during one oscillation (starting at an amplitude) are:

PE→KE→PE→KE→PE

As the mass–spring pendulum oscillates, it transfers energy between kinetic, gravitational potential and elastic potential. It is usual to consider the potential energy to be zero at equilibrium.

Graphical representation of energy transfers for an undamped oscillator

Kinetic energy (KE) $= \frac{1}{2}mv^2 = \frac{1}{2}m\omega^2 A^2 \sin^2 \omega t$

Total energy (TE) = KE + potential energy
(PE) = KEmax $= \frac{1}{2}m\omega^2 A^2$

PE = TE − KE $= \frac{1}{2}m\omega^2 A^2 (1 - \sin^2 \omega t)$

PE $= \frac{1}{2}m\omega^2 A^2 (1 - \sin^2 \omega t)$

PE $= \frac{1}{2}m\omega^2 A^2 (\cos^2 \omega t)$

 Maths skills Remember that $\sin^2 \omega t + \cos^2 \omega t = 1$

Energy transfers for an undamped simple pendulum

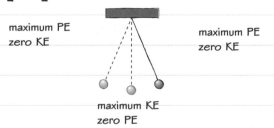

maximum PE
zero KE

maximum PE
zero KE

maximum KE
zero PE

During one oscillation, starting at an amplitude, the following transfers take place:

PE→KE→PE→KE→PE

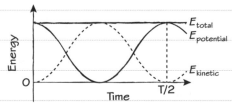

The value of E_{total} is constant because we have assumed frictional forces are zero and the oscillation is undamped.

Damped oscillations

When an oscillator moves against frictional forces, the work it does transfers energy by heating and so its total energy and amplitude both decay. This process is the **damping** of the oscillation.

During one oscillation, starting at an amplitude, the following energy transfers take place:

PE→KE→PE→KE→PE
    ↓   ↓
Work done against damping forces

When the oscillator is damped, the total energy of the system decays. Since TE $= \frac{1}{2}m\omega^2 A^2$ this results in a decaying amplitude with time:

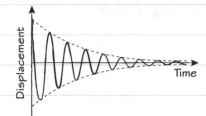

(a) Calculate the total energy in a mass–spring oscillator that has a time period of 1.2 s and a mass of 250 g when it oscillates with an amplitude of 2.0 cm. **(2 marks)**

Total energy $= \frac{1}{2}m\omega^2 A^2$

$= \frac{1}{2} \times 0.25 \times (4\pi^2/1.2^2) \times 0.020$

$= 1.4 \times 10^{-3}$ J

(b) By what factor has the total energy reduced when the amplitude falls to 1.0 cm? **(2 marks)**

TE $\propto A^2$ so if A halves, TE falls to $(\frac{1}{2})^2 = \frac{1}{4}$ of its previous value.

Now try this

Calculate the total energy of a mass–spring oscillator with an oscillator mass of 0.42 kg, a spring constant k of 25 N m^{-1} and an amplitude of 5.0 cm. **(4 marks)**

Forced oscillations and resonance

Systems that can oscillate naturally can also be forced to oscillate.

Free and forced oscillations

A **free oscillation** has no external energy source after its initial displacement.

A **forced oscillation** is driven by a periodic force that supplies energy to the oscillator.

f_0 is the **natural frequency**

f_d is the **forcing frequency**, or driving frequency

If the oscillator moves against friction, it is damped. Free and forced oscillators can be damped.

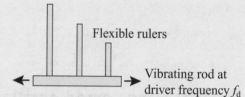

pendulum is displaced and released once, and vibrates at frequency f_0

hand vibrates at frequency f_d and pendulum vibrates at f_d

Worked example

A student is investigating how buildings respond to earthquakes. She sets up an experiment using a series of flexible rulers of different lengths (to represent buildings of different heights) attached to a rod that can be made to vibrate horizontally at different frequencies (to represent the ground).

Flexible rulers

Vibrating rod at driver frequency f_d

When the rod vibrates the rulers sway sideways with different amplitudes.

(a) Suggest a way that the student could find out the natural frequency of swaying for each ruler. **(3 marks)**

Keep the rod still. Displace the ruler and let it oscillate freely. Measure this frequency – it is the natural frequency.

(b) Suggest two factors related to the vibrations of the rod that will affect the amplitude with which any particular ruler sways. **(2 marks)**

Amplitude and frequency.

(c) Suggest, with an explanation, how the rulers will respond as she increases the frequency from a low value to a high value. **(5 marks)**

At very low frequencies all the rulers will move back and forth with the same amplitude and frequency as the driver.

As the frequency increases the amplitude of the longest ruler will increase to a maximum. This is resonance.

The shorter rulers will resonate at higher frequencies, the medium one next and then the shortest one.

At very high frequencies the rulers will all have very small amplitudes because they cannot respond quickly enough to the rapid changes.

(d) Suggest why, when an earthquake strikes a city, some buildings collapse while others nearby do not. **(4 marks)**

Those that collapse might have a natural frequency close to the driving frequency from the seismic waves. The building will respond strongly (resonate) and the large amplitude motion will destroy its structure.

Resonance

When the heavy driver is set in motion, all the light pendulums are forced to oscillate by the string that connects them. Pendulum C, with the same length as the driver, responds most strongly and has a large amplitude. This is called **resonance**.

Resonance occurs when the forcing frequency is equal to the natural frequency of the forced oscillator.

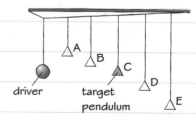

driver

target pendulum

Now try this

Earthquakes can devastate towns and cities, but not all buildings are equally affected. Suggest and explain three reasons why one building might collapse whilst another building adjacent to the first might not. **(6 marks)**

Driven oscillators

Systems that are forced to oscillate will resonate in the right conditions.

The response of a driven oscillator

At the resonance frequency, the forcing oscillator continually transfers energy to the forced oscillator. Its amplitude grows until the rate at which energy is supplied is balanced by the rate at which energy is lost to the surroundings by work done against damping forces.

In mechanical systems damping might occur because of the plastic deformation of vibrating objects – this will limit the amplitude of resonance.

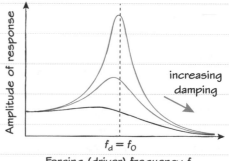

If there is no damping, the oscillator continues to absorb energy until it is destroyed.
As damping increases, the resonance becomes less sharp and the natural frequency is slightly reduced.

Investigating forced oscillations and resonance

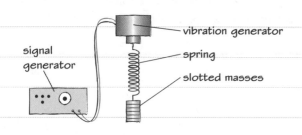

A system like the one above can be used to determine an unknown mass. Known masses are used first to determine the spring constant. Then the unknown mass is substituted. The forcing frequency is varied until the oscillator resonates. At resonance:

$$f_d = f_0 = \frac{1}{2\pi} \sqrt{\left(\frac{k}{m}\right)}$$

Worked example

The rear view mirror of a car vibrates violently at certain speeds.

(a) Explain why this happens. **(4 marks)**

This is a resonance effect. The mirror vibration is being driven by the vibrations of the car engine. At a certain speed the forcing vibrations have a frequency equal, or very close, to the natural frequency of vibration of the mirror so it resonates and vibrates with large amplitude.

(b) Suggest two ways in which the effect could be removed or prevented. **(2 marks)**

One way would be to increase the damping. This could be done by changing its mounting, e.g. adding a plastic support. Another way would be to change its natural frequency of vibration. This could be done by changing its mass or changing the dimensions of its support.

Now try this

The wheels on a car are supported by suspension springs.

(a) If the car goes over a bump in the road what is the effect on this mass–spring system? **(2 marks)**

(b) The suspension system also includes shock absorbers. Explain the purpose of these devices. **(2 marks)**

Exam skills 13

This exam-style question uses knowledge and skills you have already revised. Have a look at pages 128–134, for a reminder about simple harmonic motion.

Worked example

A car suspension system can be modelled as a damped mass–spring oscillator, as shown in the diagram.

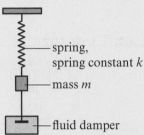

spring, spring constant k

mass m

fluid damper

The fluid damper provides a frictional force that opposes the motion of the oscillator.

(a) (i) When the car goes over a bump in the road it oscillates vertically with a time period of 0.80 s and an initial amplitude of 10 cm. The car, with driver, has a total mass of 1250 kg. Calculate the effective spring constant of the car's suspension. **(3 marks)**

For a mass–spring system $T = 2\pi\sqrt{\left(\dfrac{m}{k}\right)}$

so $k = \dfrac{4\pi^2 m}{T^2} = 77\,000\,\text{N m}^{-1}$ (2 s.f.)

A question like this one places physics in context, in this case modelling a car's suspension. You are not expected to know anything in detail about how cars are constructed or how their suspension is configured. You need to apply your knowledge of the important underlying physics – mass–spring systems, damping, forced oscillations and resonance.

Answers should be rounded to an appropriate number of significant figures. The data used in the calculation is to a minimum of 2 s.f. – 10 cm and 0.80 s – so this answer has been rounded to 2 s.f. as well.

(ii) Sketch a graph to show how the displacement of the car varies with time from the moment it first reaches the bump in the road. Indicate approximate values on the axes. **(3 marks)**

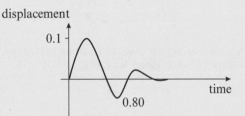

displacement

0.1

time

0.80

The important point here is to show that the oscillation is damped. The amount of damping doesn't matter, just make sure the amplitude decays. You are also asked to indicate values on the axes. The maximum displacement is 10 cm and the period of oscillation is 0.80 s so these should be marked onto the axes. Make sure that the time period is shown after one complete cycle or mark 0.40 s after half a cycle.

(iii) Calculate the maximum energy stored in the oscillation after the car hits the bump. **(2 marks)**

For simple harmonic motion: $v = -\omega A \sin\omega t$

Kinetic energy $E_K = \tfrac{1}{2}mv^2 = \tfrac{1}{2}m\omega^2 A^2 \sin^2 \omega t$

$E_{Kmax} = \tfrac{1}{2}m\omega^2 A^2 = \tfrac{1}{2}m(2\pi f)^2 A^2 = 386\,\text{J}$

Notice how the working has been shown in logical steps, starting with the equation for velocity and the definition of kinetic energy.

The calculation here follows from the explanation in part (i). This has also been rounded to 2 s.f.

(iv) Explain what happens to this energy. **(2 marks)**

The oscillator does work against viscous damping forces, transferring the kinetic energy of the oscillator into thermal energy in the fluid.

Command word: Explain

If a question asks you to **explain** why something happens you should:

Write in full sentences.

Use correct scientific language. ✓

Answers

1. SI units

1 Force has base units kg m s^{-1} (from $F = ma$); s, the distance through which the force is applied in the line of action of the force, has the base unit m ∴ the base unit of work is $\text{kg m s}^{-1} \times \text{m} \rightarrow \text{kg m}^2 \text{s}^{-1}$.

2 mv^2 has base units $(\text{kg}) \times (\text{m s}^{-1})^2 \rightarrow (\text{kg}) \times (\text{m}^2 \text{s}^{-2}) \rightarrow \text{kg m}^2 \text{s}^{-1}$.
mgh has base units $(\text{kg}) \times (\text{m s}^{-2}) \times (\text{m}) \rightarrow \text{kg m}^2 \text{s}^{-2}$.
The two sets of base units are the same.

3 $s = \frac{1}{2}(u + v)t$ has units (left-hand side) s in m; (right-hand side) $(u + v)$ in $\text{m s}^{-1} \times t$ in s = m.

2. Practical skills

1 (a) Human reaction time. The stopwatch may not be started and stopped precisely at the start and finish of the period being timed. This is a random error.
(b) By repeating the measurement several times and taking the average value.
(c) As human reaction time is typically around 0.2–0.3 seconds, its effect on the uncertainty of short time intervals is much greater.

2 Graphs will show trends in results. Your results may produce a graph line which is straight or a smooth curve. Results that stand out from the trend could be mistakes in measurement and should be repeated if possible.

3 (a) A long tape measure is preferable to using a metre rule, as every time you move the metre rule there is a possibility of extra errors.
(b) A metre rule.
(c) A micrometer or digital calipers. A higher degree of precision is essential when measuring small quantities.

3. Estimation

1 car: 10^3 kg, atom: 10^{-27} kg, planet: 10^{24} kg

2 Estimate the mass of an elephant by comparing with a human being, say mass 5000 kg, weight 50 000 N; estimate area of elephant feet, say $4 \times 0.5 \text{ m} \times 0.5 \text{ m} \sim 1 \text{ m}^2$, so pressure $\sim 50\,000 \text{ N m}^{-2}$.
Estimated mass of passenger aircraft: hollow metal tube floats on water, so $\sim 50 \text{ m} \times 5 \text{ m} \times 5 \text{ m} \times 500 \text{ kg m}^{-3}$ $\sim 500\,000 \text{ kg}$, area of wheels in contact with ground $\sim 50 \text{ m}^2$, pressure $\sim 100\,000 \text{ N m}^{-2}$.
In fact large passenger aircraft have mass $\sim 300\,000 \text{ kg}$; aircraft tyres may be inflated to $\sim 1\,400\,000 \text{ N m}^{-2}$.

4. SUVAT equations

1 (a) $u = 20 \text{ m s}^{-1}$, $t = 10 \text{ s}$, $v = 30 \text{ m s}^{-1}$, $s = ? \text{ m}$
$$s = \frac{(u + v)t}{2} = \frac{(20 + 30) \times 10}{2} = 250 \text{ m}$$
(b) $v = u + at$
$$a = \frac{(v - u)}{t} = \frac{(30 - 20)}{10} = 1.0 \text{ m s}^{-2}$$

2 (a) $u = 80 \text{ m s}^{-1}$, $v = 0 \text{ m s}^{-1}$, $a = -9.81 \text{ m s}^{-2}$, $s = ? \text{ m}$
$v^2 = u^2 + 2as$
$= (-80)^2 - 2 \times 9.81 \times s = 0$
$$s = \frac{6400}{(2 \times 9.81)} = 326 \text{ m}$$
(b) $s = ut + \frac{1}{2}at^2$
$-326 = 0 - \frac{1}{2} \times 9.81 \times t^2$
$t^2 = 66.5$
$t = 8.2 \text{ s}$

5. Displacement–time, velocity–time and acceleration–time graphs

1 (a) velocity, (b) displacement, (c) acceleration.

2 A C D E
(i)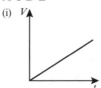

6. Scalars and vectors

1 Air pressure is a scalar quantity, because it has magnitude but no direction. It acts equally in all directions.

2 (a) $\text{average speed} = \frac{\text{distance}}{\text{time}} = \frac{400}{545.9} = 7.2 \text{ m s}^{-1}$
(b) $\text{average velocity} = \frac{\text{displacement}}{\text{time}} = 0 \text{ m s}^{-1}$

7. Resolution of vectors

(a) $T_{\text{Ah}} = 87.5 \cos 52° = 53.9 \text{ N left}$
$T_{\text{Av}} = 87.5 \sin 52° = 69.0 \text{ N upwards}$
$T_{\text{Bh}} = 62.5 \cos 31° = 53.6 \text{ N right}$
$T_{\text{Bv}} = 62.5 \sin 31° = 32.2 \text{ N upwards}$

(b) The horizontal components are (very nearly) equal and opposite; they cancel each other out, so there is no horizontal resultant force on the painting.

8. Adding vectors

1 A scale drawing will give the resultant force on the pendulum bob as **6 N** acting **horizontally to the right**.

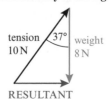

2 A scale drawing will give the resultant velocity of the boat of **5.8 m s^{-1}** at an angle of **N 44.3° E.**

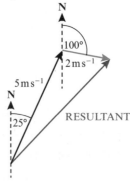

9. Projectiles

(a)

vertical component of velocity $= 200 \sin 75° \text{ m s}^{-1}$
$= 193 \text{ m s}^{-1}$

horizontal component of velocity $= 200 \cos 75° \text{ m s}^{-1}$
$= 52 \text{ m s}^{-1}$

(b) The initial vertical component of the velocity $u = 193\,\text{m s}^{-1}$.

> The ball climbs until all its initial kinetic energy is transferred to gravitational potential energy, coming momentarily to rest at the top of its flight. It then falls back to Earth and all the GPE is converted back to KE by the time it returns to ground level. Assume that we can ignore the effect of air resistance.

This means that when the ball returns to ground level it has the same vertical speed that it started with, but now downwards instead of upwards: $v = -193\,\text{m s}^{-1}$
We know that $g = -9.81\,\text{m s}^{-2}$. The minus means downwards.
$v = u + at$ so $t = \dfrac{(-193 - 193)}{(-9.81)} = 39.3\,\text{s}$

(c) The horizontal range = horizontal component of velocity × time of flight
Again, we assume that there is negligible air resistance and the horizontal velocity component remains constant throughout the flight.
horizontal range = $52\,\text{m s}^{-1} \times 39.3\,\text{s} = 2040\,\text{m}$

10. Free body diagrams

1

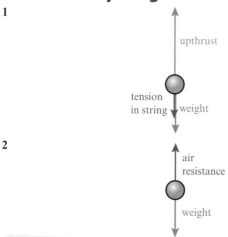

2

air
resistance

weight

> Note that air resistance is speed dependent. At some speed, air resistance and weight will be equal in magnitude.
> Upthrust due to displaced air is omitted in this case because the size of this force will be insignificant compared with the other two forces.

3

normal reaction   normal reaction

weight

> The pair of normal reaction forces acting on the sphere will be of the same magnitude, with mirrored directions.

11. Newton's first and second laws of motion

1   $600\,\text{g} = 0.600\,\text{kg}$, $F = ma$ so $F = 0.600 \times 5.0 = 3.0\,\text{N}$

2   $F = ma$, $a = \dfrac{F}{m} = \dfrac{125}{50} = 2.5\,\text{m s}^{-2}$

3   (a) $m = \dfrac{F}{a} = \dfrac{180\,000}{1.8} = 100\,000\,\text{kg}$

(b) As the plane speed increases, so does the size of the opposing drag force caused by air resistance. When the drag force is equal and opposite to the thrust of the engines, the resultant horizontal force is zero. From $F = ma$ this means acceleration is now zero and the plane will continue in steady horizontal flight, as described in Newton's first law, at its maximum speed.

(c) If the plane mass is smaller it will be able to accelerate at a greater rate, for the same force.

12. Measuring the acceleration of free fall

1   (a) $W = mg = 55 \times 9.81 = 540\,\text{N}$

(b) The gravitational field strength at the surface of the Moon is much smaller than that at the surface of the Earth, so the springs in the scale do not compress as much, so mass readings are smaller. Of course, the mass has not changed, but the force per kilogram, gravitational field strength on the Moon is $\dfrac{9.11}{55}$ of the Earth's gravitational field strength, about one-sixth as strong.

2   (a) Measure from the bottom of the steel ball to the surface of the trapdoor switch; ensure measurements are perpendicular to the metre rule by using a set square.

(b) Using a millisecond timer is necessary because the time intervals measured are quite short; operating the clock electronically removes the potentially large errors due to human reaction times; repeating the time measurements at each height and taking an average reduces the effect of random errors.

(c) Using a graph improves the accuracy of the result as it averages all the values found. It will help to identify erroneous readings, as they will stand out if not close to the line of best fit. The effect of errors in timing has a bigger effect on the overall accuracy of the calculated value because the values for t are squared.

13. Newton's third law of motion

1   (a) Three forces act on the balloon (neglecting the negligible upward force on the balloon of the surrounding air):
   • its weight acting downwards – a gravitational force
   • the tension force in the nylon string, acting at an angle to the vertical
   • the electric force of repulsion between its charge and the charge on the other balloon, acting horizontally to the left.

(b) The reaction to the weight of the balloon is the upward gravitational pull of the balloon on the Earth, equal in magnitude, opposite in direction.
   • The tension force in the string is pulling on the support shown at the top of the diagram; it acts along the string, equal in magnitude to the pull on the balloon but opposite in direction.
   • The reaction to the electric force is the electric force on the right-hand balloon; it has the same magnitude and line of action but acts horizontally to the right.

2   Two significant forces act on the parachutist:
   • Weight – the gravitational force. This may be considered to be constant throughout the fall. The reaction to this force is the equal and opposite gravitational pull of the parachutist on the Earth – the Earth, being considerably more massive than the parachutist, is virtually unaffected by this upward pull.
   • Air resistance or drag acting upwards on the parachutist. This force is speed dependent, so increases from zero when the parachutist first jumps and increases with her speed. The reaction to this is the downward force on the air as the parachutist 'pushes' her way through it.

(When the weight force exceeds the upward force of air resistance, the parachutist accelerates. Eventually the two forces are in equilibrium and the parachutist is travelling at a terminal velocity. This is somewhere between 150 and 200 km h^{-1} for human beings. When the parachute opens, the upward force of air resistance on the parachute increases considerably, and, via the tension in the parachute cords, causes a net upward force on the parachutist. In videos of parachutists it may appear that the parachutist shoots upwards at this point, but this is an optical illusion. In fact, the parachutist carries on falling but is decelerating. Deceleration continues until, at a much lower downward terminal velocity, the two forces acting on the parachutist are in equilibrium.)

14. Momentum

1 total momentum before = total momentum after
$m_A u_A + m_B u_B + m_B v_B$
$= m_2 v_2 = (4.0 \times 1.0) + (2.0 \times 3.0) = (4.0 \times 2.0) + (2.0 \times v_B)$
$\therefore v_B = \dfrac{(10.0 - 8.0)}{2.0} = 1.0\,\text{m s}^{-1}$ in the same direction as before the collision.

2 Assuming negligible friction, and applying the law of conservation of momentum:
$m_1 u_1 + m_2 u_2 = m_1 v_1 + m_2 v_2$
$\therefore (200 \times 6.5) + (m \times 0) = (200 + m) \times 2.6$
$\therefore m = \dfrac{1300}{2.6} - 200 = 300\,\text{kg}$

15. Moment of a force

(a)
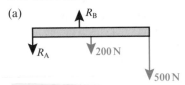

The free body diagram shows only the body we are concerned with, in this case the diving board, and the forces that act on the diving board. We assume that the board is uniform, therefore, the weight of the board, 200 N, acts through the centre of gravity in the middle of the board. The support at B exerts an upward force on the board, R_B, and the support at A exerts a downward force, R_A. If R_A did not act downwards, the board could not be in equilibrium as there would be a net clockwise moment about B.

(b)
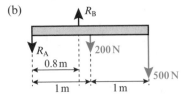

Taking moments about B means that we do not need to know the force acting through B.
For rotational equilibrium,
clockwise moments about B = anticlockwise moments about B
$(500 \times 1.2) + (200 \times 0.2) = R_A \times 0.8$
$R_A = 800\,\text{N}$ acting downwards (the support at A is in tension).

(c) For rotational equilibrium,
clockwise moments about A = anticlockwise moments about A
$(500 \times 2) + (200 \times 1) = R_B \times 0.8$
$R_B = 1500\,\text{N}$ acting upwards (the support at B is in compression)
We can verify this answer by checking that the resultant force vertically on the board is zero:
force upward = force downward
$1500 = 800 + 200 + 500$

17. Work

1 $\Delta W = F \Delta s = 6000 \times 200 = 1.2 \times 10^6\,\text{J}$
2 force in line of motion is 700 000 N
distance travelled 800 000 m h^{-1} × 3 h
$\therefore$ work done = $700\,000 \times 2\,400\,000 = 1.68 \times 10^{12}\,\text{J}$
3 force down the plane (line of motion) is $F = 5.0 \times \sin 20°\,\text{N}$
$\Delta W = F \Delta s = 5.0 \times \sin 20° \times 1.5 = 2.6\,\text{J}$

18. Kinetic energy and gravitational potential energy

1 (a) The vertical height lost is $1.50 \times \sin 37° = 0.90\,\text{m}$
$\therefore \Delta E_{grav} = mgh = 1.2 \times 9.81 \times 0.90 = 10.6\,\text{J} = 11.0\,\text{J}$ to 2 s.f.

(b) $\Delta E_K = \frac{1}{2}mv^2 = 10.6\,\text{J} \therefore v = \sqrt{\left(\dfrac{2 \times 10.6}{1.2}\right)} = 4.2\,\text{m s}^{-1}$ to 2 s.f.

2 Convert 108 km h^{-1} to m s^{-1}: 30 m s^{-1}
$\Delta E_K = \frac{1}{2}mv^2 = 0.5 \times 1200 \times (30)^2 = 540\,000\,\text{J}$

3 Assuming no energy is lost through air resistance, all the initial kinetic energy is converted to gravitational potential energy when the firework comes to a momentary halt at the top of its flight.
E_k on launch = ΔE_{grav} at top of flight
$\frac{1}{2}mv^2 = mgh$
$h = \dfrac{v^2}{2g} = 81.5\,\text{m}$

19. Conservation of energy

(a) The total KE of the stone is $\frac{1}{2} \times 0.05 \times (50)^2 = 62.5\,\text{J}$.
(b) At the top of the stone's flight, the stone will have gained gravitational potential energy, but it will still have kinetic energy because its horizontal velocity component is unchanged, if there is no air resistance.
(c) The change in height is Δh, so:
$mg\Delta h = \frac{1}{2}mv^2 \rightarrow \Delta h = \dfrac{v^2}{2g}$
v is the vertical component of the initial velocity of the stone, $50 \sin 53° = 39.93\,\text{m s}^{-1}$
The height of the stone's flight $\Delta h = \dfrac{(39.93)^2}{(2 \times 9.81)} = 81.3\,\text{m}$

20. Work and power

(a) $\Delta E_{grav} = mg\Delta h = 45 \times 9.81 \times 0.2 \times 30 = 2648.7$ or 2600 J to 2 s.f.
(b) This is equal to the work done by the student moving against the force of gravity.
$P = \dfrac{\Delta E_{grav}}{t} = \dfrac{2648.7}{15} = 180\,\text{W}$ to 2 s.f.

21. Efficiency

1 The hoist must exert an upward force on the load equal to its weight. In one second it moves the load through 1.5 m
$\therefore$ it does work $W = F \Delta s = 1200 \times 1.5 = 1800\,\text{J}$ in one second, so the useful output power is 1800 W or 1.8 kW.
efficiency = $\dfrac{1.8\,\text{kW}}{3.0\,\text{kW}} = 0.6$.

2 (a) In one second, the car travels 27 m, so useful energy output per second = $F \Delta s = 400 \times 27 = 10\,800\,\text{W}$.
(b) The useful power output is 20% or 0.2 of the total power input
$\therefore$ total power input = $\dfrac{10\,800 \times 100}{20} = 54\,000\,\text{W}$ (54 kW)
(c) The wasted energy/power is mainly converted into heat and some sound.

23. Basic electrical quantities

1 (a) $I = \dfrac{P}{V} = \dfrac{60}{230} = 0.26\,\text{A}$
(b) $Q = It = 0.26 \times (5.0 \times 60) = 78\,\text{C}$
2 (a) $Q = It = 0.25 \times 1800 = 450\,\text{C}$
(b) $P = \dfrac{E}{t} = \dfrac{2400}{1800} = 1.3\,\text{W}$
(c) $V = \dfrac{E}{Q} = \dfrac{2400}{450} = 5.3\,\text{V}$

24. Ohm's law

(a) R at 0.1 V is $10\,\Omega$; R at 6 V is $100\,\Omega$

(b) The heating effect of 10 mA is quite small. When 6 V is applied to the lamp at room temperature, the current is much larger. More charge per second passes through the lamp, transferring more energy per second and so increasing the temperature of the lamp. As the lamp's temperature increases, the positive ions in the wire that make up the filament vibrate more, making it harder for the charge-carrying electrons to pass. This increases the resistance of the lamp.

25. Conservation laws in electrical circuits

1 The total current out of any point in a circuit must **equal** the **total** current into that point as **charge** must be conserved.

2 The p.d.s across components in series must **add up/sum** to the supply e.m.f. The p.d.s across components in parallel must be **the same**. The total **energy** supplied by a battery per coulomb of charge circulated must be **equal to** the total energy transferred in the circuit.

26. Resistors

(a) $\dfrac{1}{R_{\text{parallel}}} = \dfrac{1}{6} + \dfrac{1}{4} = \dfrac{5}{12}\,(\text{k}\Omega)^{-1}$

$R_{\text{parallel}} = 2.4\,\text{k}\Omega$

$R_{\text{total}} = 600\,\Omega + 2400\,\Omega = 3000\,\Omega$ or $3.0\,\text{k}\Omega$

(b) $R_{\text{series}} = 12\,\Omega + 48\,\Omega = 60\,\Omega$

$\dfrac{1}{R_{\text{parallel}}} = \dfrac{1}{60} + \dfrac{1}{90} = \dfrac{5}{180}\,\Omega^{-1}$

$R_{\text{parallel}} = 36\,\Omega$

27. Resistivity

1 **A** $9\,\Omega$

Tripling the length increases the resistance $\times 3$.

Doubling the thickness increases the cross-sectional area $\times 4$.

Doing both therefore changes $R \times \frac{3}{4}$

2 $A = \pi r^2 = \pi \times \left(\dfrac{0.40}{2 \times 10^{-3}}\right)^2 \text{m}^2$

$R = \dfrac{\rho l}{A} = \dfrac{1.7 \times 10^{-8} \times 3.5}{\left[\pi \times \left(\dfrac{0.40 \times 10^{-3}}{2}\right)^2\right]} = 0.47\,\Omega$

28. Resistivity measurement

(a) Resistivity = gradient × area A

Area $A = 0.2\,\text{mm}^2 = 2 \times 10^{-7}\,\text{m}^2$

Gradient measurement should yield a value of $\approx 1.1 \times 10^{-6}\,\Omega\,\text{m}$

(b) This resistivity is much higher than that of copper wires. Nichrome is suitable for use in heating elements because its high resistance causes it to heat up easily.

29. Current equation

1 B, half that in wire 1

2 The temperature of both wires is the same.

30. E.m.f. and internal resistance

(a) The internal resistors combine according to the parallel rule, so this cell has an internal resistance of $2.5\,\Omega$. The e.m.f. of these identical cells in parallel is 2 V. Current drawn $= \frac{2}{17.5} = 0.11\,\text{A}$

terminal p.d. $= 2 - (0.11 \times 2.5) = -1.7\,\text{V}$

(b) (i) The e.m.f. in series is 4 V and the internal resistance is $10\,\Omega$.

(ii) $I = \dfrac{4}{(10 + 15)} = 0.16\,\text{A}$

terminal p.d. $= 4 - (0.16 \times 10) = 2.4\,\text{V}$

31. Potential divider circuits

(a) In bright light:

$V_{\text{out}} = \dfrac{(5 \times 25)}{(25 + 50)} = \dfrac{125}{75} = 1.7\,\text{V}$

In the dark:

$V_{\text{out}} = \dfrac{(5 \times 25)}{(25 + 2500)} = \dfrac{125}{2525} = 0.05\,\text{V}$

Range of output p.d. = 0.0(5)−1.7 V

(b) By similar calculations, range of output p.d. = 0.8−4.5 V

(c) By similar calculations, range of output p.d. = 4.0−5.0 V

33. Density and flotation

(a) (i) $0.6\,\text{g cm}^{-3}$ (ii) $600\,\text{kg m}^{-3}$

(b) (i) It will float because its density is less than that of water.

(ii) The volume of the block is $400\,\text{cm}^3$ and it must displace a volume of water whose mass is equivalent to 240 g,

volume $= \dfrac{\text{mass}}{\text{density}}$ therefore volume of water displaced

$= \dfrac{240\,\text{g}}{1\,\text{g cm}^{-3}} = 240\,\text{cm}^3$, therefore $160\,\text{cm}^3$ of block remains above the water line = 40%

34. Viscous drag

1 (a) (i) $V = \frac{4}{3}\pi r^3 = 1.44 \times 10^{-6}\,\text{m}^3$

(ii) $m = \rho V = 3.6 \times 10^{-3}\,\text{kg}$

(iii) $W = mg = 3.5 \times 10^{-2}\,\text{N}$

(b) Upthrust $= (\rho_g - \rho_l)Vg = 1300 \times 1.44 \times 10^{-6} \times 9.81$

$= 0.018\,\text{N}$

(c) Drag $F = 6\pi\eta rv = 6 \times \pi \times 0.5 \times 0.007 \times 0.10$

$= 6.6 \times 10^{-3}\,\text{N}$

2 The coefficient of viscosity will increase, which will cause the drag to increase, so the sphere will slow down to a lower value of terminal velocity.

35. Hooke's law

1 The limit of proportionality is the maximum force that may be applied to a wire before it stops obeying Hooke's law. The elastic limit is the maximum force that can be applied to the wire with it still returning to its original length when the force is removed. For some materials, these two points may be the same.

2 Each spring in this combination will be subject to the same load force of 20 N, so the $k = 100\,\text{N m}^{-1}$ spring will stretch 0.20 m and the $k = 200\,\text{N m}^{-1}$ spring will stretch 0.10 m, so total extension = 0.30 m.

The $k = 100\,\text{N m}^{-1}$ spring will store $\frac{1}{2} \times 20 \times 0.20 = 2.0\,\text{J}$ of elastic strain energy and the $k = 200\,\text{N m}^{-1}$ spring will store $\frac{1}{2} \times 20 \times 0.10 = 1.0\,\text{J}$ of elastic strain energy.

36. Young modulus

(a) $A = \pi \times \left(\dfrac{18 \times 10^{-3}}{2}\right)^2 \text{m}^2$

stress $\sigma = \dfrac{F}{A} = \dfrac{10 \times 10^3}{(\pi \times 81 \times 10^{-6})} = 39.3\,\text{MPa}$

(b) $E = \dfrac{\sigma}{\varepsilon}$ so $\varepsilon = \dfrac{\sigma}{E}$ and thus extension $\Delta x = \dfrac{\sigma}{E}x$.

Substituting values: $\Delta x = \dfrac{39.3 \times 10^6}{2 \times 10^{11}} \times 15$

$= 2.95 \times 10^{-3}\,\text{m}$ (2.95 mm)

(c) $\varepsilon = \dfrac{\Delta x}{x}$ therefore $\varepsilon = \dfrac{2.95 \times 10^{-3}}{15}$

$= 1.96 \times 10^{-4}$. Remember, no units.

(d) $\sigma_{\text{max}} = \dfrac{F_{\text{max}}}{A}$ therefore $F_{\text{max}} = \sigma_{\text{max}} \times A$

$= 400 \times 10^9 \times (\pi \times 81 \times 10^{-6})$

$= 1.02 \times 10^8\,\text{N}$

(e) $\sigma_{max} = \dfrac{10 \times 10^3 \,\text{N}}{A_{min}}$ therefore $A_{min} = \dfrac{10 \times 10^3 \,\text{N}}{\sigma_{max}}$

$A_{min} = \dfrac{10 \times 10^3}{400 \times 10^9} = 2.50 \times 10^{-8} \,\text{m}^2$

$A = \pi r^2$

hence, minimum diameter $2r = 1.78 \times 10^{-4} \,\text{m}$

38. Waves

(a) Wavelength λ can be measured directly from displacement–distance graph: $\lambda = 0.20 \,\text{m}$
(b) In 0.20 s the wave crest has travelled 5 cm to the right, therefore speed v is $0.25 \,\text{m s}^{-1}$.
(c) $f = \dfrac{v}{\lambda} = \dfrac{0.25}{0.2} = 1.25 \,\text{Hz}$

39. Longitudinal and transverse waves

(a) longitudinal
(b) not waves
(c) transverse
(d) longitudinal
(e) transverse
(f) transverse
(g) not waves
(h) longitudinal

40. Standing waves

1 $v = f\lambda = 400 \times 0.85 = 340 \,\text{m s}^{-1}$.
2 (a)

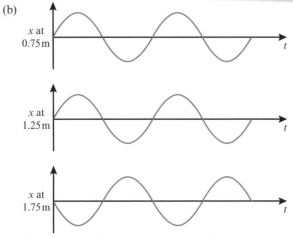

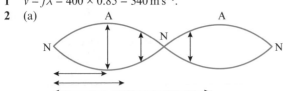

The horizontal and vertical arrows are not required for this part of the question, but are used for constructing the answers to part (b).

(b)

x at 0.75 m

x at 1.25 m

x at 1.75 m

Notice that the first two points are in phase, as they are both within the same pair of antinodes.
Amplitude at 1.25 m is smaller than at 0.75 m (which is an antinode).
At 1.75 m (between node 2 and node 3) the oscillation is 180° out of phase (antiphase) with the first two.

41. Phase and phase difference

(a) 0–A: 360°, 2π rad; A–B: 180°, π rad; B–C: 270°, $\frac{3}{2}\pi$ rad.
(b) The green wave is $\frac{1}{4}$ of a period T out of phase with the blue wave – the green wave leads the blue by 90° $\left(\frac{\pi}{2}\,\text{rad}\right)$.

42. Superposition and interference

In the example, the path difference from the two sources to X was a whole number of wavelengths, so the waves were in phase when they arrived at X and therefore interfered constructively. The result was a bigger amplitude disturbance at X.
If the two sources are in antiphase, the waves arriving at X will be in antiphase (half a wavelength out of phase) so they will now superpose and interfere destructively. The result is that there will be almost no disturbance in the water surface at X.

43. Velocity of transverse waves on strings

1 Rearrange the equation to give $\mu = \dfrac{T}{(2lf)^2}$
mass per unit length of the bottom E string
$\mu = \dfrac{75}{(2 \times 82.4 \times 0.64)^2} = 0.0067 \,\text{kg m}^{-1}$
2 Since $f \alpha \dfrac{1}{l}, \dfrac{f_1}{f_2} = \dfrac{l_2}{l_1}$ so $\dfrac{82.4 \,\text{Hz}}{110 \,\text{Hz}} = \dfrac{l_2}{0.64 \,\text{m}} \rightarrow l_2 = 0.48 \,\text{m}$.

44. The behaviour of waves at an interface

(a) The time between the transmitted and reflected pulse is $5 \times 20 \,\mu\text{s}$, that is, $1.0 \times 10^{-4} \,\text{s}$. Radar waves travel at light speed, $3.00 \times 10^8 \,\text{m s}^{-1}$, and therefore the distance travelled out to the object and back ($2d$) is:
$2d = ct = (3.00 \times 10^8) \times 1.0 \times 10^{-4} = 30\,000 \,\text{m}$ or 30 km
Therefore, the distance d to the object is 15 km.
(b) Radar uses narrow beams, so the usual explanation involving the inverse square law is not applicable here. Depending on the size and shape of the detected object, only a proportion of the beam energy will be reflected; there will also be signal loss due to scattering.

45. Refraction of light and intensity of radiation

1 $n_w = \dfrac{c}{v_w}$ so v_w, the speed of light in water, is given by
$v_w = \dfrac{c}{n_w} = \dfrac{3.00 \times 10^8}{1.33} = 2.26 \times 10^8 \,\text{m s}^{-1}$
2 The surface area of a sphere is proportional to the radius squared. This means that a sphere of radius $10R$ has a surface area $100 \times$ larger than a sphere of radius R. Therefore the intensity of the radiation reaching the planet Saturn is $0.01 \times$ of that reaching the Earth $\rightarrow 14 \,\text{W m}^{-2}$

46. Total internal reflection

1 B

The larger the value of n the more slowly light travels in the medium, and light must travel faster in the cladding glass than in the core glass. D suggests a medium in which light travels faster than in a vacuum. This is not possible.

2 The speed of sound in plastic must be greater than the speed of sound in air.

48. Lenses and ray diagrams

1 The image is formed 60 cm from the lens. It is inverted, magnified and real. Film and projectors use this arrangement to project real magnified images onto a screen. In practice, the object, the picture on the film, is placed much closer to the focal point as this produces much greater magnification and forms an image on a screen much further from the projector.

2 The rays emerging from the lens are diverging and appear to come from a point on the same side of the lens as the object, O. The image, I, formed is **virtual** as it is not the source of real rays of light. The image is also upright, magnified and 30 cm from the plane of the lens. Application: lens used as a magnifying glass.

> Rays 1 and 2 are sufficient; ray 3 is included to show how it is drawn.

49. Lens formula

1 $v = -0.24$ cm as the image is virtual.
$$\frac{1}{f} = \frac{1}{u} + \frac{1}{v} \rightarrow \frac{1}{0.08} = \frac{1}{u} - \frac{1}{0.24}$$
$u = 0.06$ m
Magnification $m = \frac{v}{u} = \frac{0.24}{0.06} = 4$

> Since only the magnitude of magnification is required, the sign of the image distance can be ignored.

2 $\frac{1}{f} = \frac{1}{u} + \frac{1}{v} \rightarrow \frac{1}{0.10} = \frac{1}{0.20} + \frac{1}{v}$
$v = 0.20$ m
$m = \frac{v}{u} = \frac{0.20}{0.20} = 1$

> This is a special case: objects at 2f are not magnified. You can prove this by calculating v when $m = 1$: $v = u$, so $\frac{1}{f} = \frac{2}{u}$, $u = 2f$

50. Plane polarisation

1 The light reflected from surfaces is partly plane polarised, so if the plane of polarisation of the filters in sunglasses cuts out horizontally polarised light it will filter out a lot of the reflected light from a surface.
If you rotate sunglasses through 90° the glare is not filtered out.

2 The light we receive directly from the Sun is not polarised, but light scattered toward the Earth by particles in the atmosphere is. When the Sun is low in the sky, the proportion of scattered polarised light reaching the Earth is greater and the stress patterns in toughened glass are therefore easier to see.

52. Diffraction grating and wavelength of light

1 $n\lambda = d \sin \theta$, so $\sin \theta = \frac{n\lambda}{d}$. The limiting value of $\sin \theta$ is 1.
θ cannot exceed 90°. Thus $\frac{n\lambda}{d} \leqslant 1 \rightarrow n \leqslant \frac{d}{\lambda}$
$d = \frac{1}{(6 \times 10^5)}$ m $\rightarrow 1.67 \times 10^{-6}$ m
$n \leqslant \frac{1.67 \times 10^{-6}}{7 \times 10^{-7}} \rightarrow n \leqslant 2.39$. Thus, the largest order is 2.

2 (a) $BP = d \sin \theta_3$, $CQ = 2d \sin \theta_3$
(b) $BP = 3\lambda$, $CQ = 6\lambda$, because the path difference between each of the rays contributing to a third-order maximum must be 3λ. This means that secondary wavelets from all slits in the diffraction grating will be in phase at the angle and a third maximum will be observed at the angle θ_3.

53. Electron diffraction

1 $\lambda = \frac{h}{p} = \frac{6.63 \times 10^{-34}}{(9.1 \times 10^{-31} \times 1.2 \times 10^5)} = 0.607 \times 10^{-8}$ m
2 Both have a single charge so both gain the same amount of energy, in eV. However, $eV = \frac{1}{2}mv^2 = \frac{p^2}{2m}$, where $p = mv$, momentum. Rearranging this gives $p = \sqrt{(2meV)}$, so the

proton will have greater momentum and therefore shorter wavelength, from $l = \frac{h}{p}$.

54. Waves and particles

$\lambda = \frac{h}{p}$
$\lambda = \frac{6.63 \times 10^{-34}}{0.163 \times 40} = 1.02 \times 10^{-34}$ m.
This wavelength is so small that the wave properties of a cricket ball have no effect on play. It will not be diffracted by anything on a human scale.

55. The photoelectric effect

1 $hf \geqslant \phi$ so $f \geqslant \frac{\phi}{h}$
Therefore, the minimum frequency of light required is $\frac{3.65 \times 10^{-19}}{6.63 \times 10^{-34}}$
$f_{min} = 5.50 \times 10^{14}$ Hz.
The maximum wavelength $\lambda_{max} = \frac{c}{f_{min}} = \frac{3.00 \times 10^8}{5.5 \times 10^{14}}$
$= 5.45 \times 10^{-7}$ m or 545 nm. This is at the violet end of the visible spectrum.

2 (a) In terms of the quantum theory of light, more intense light, that is, more energy arriving per second per m², means more photons arriving at the surface of the metal per second. Each photon liberates one electron. Therefore greater light intensity means more photoelectrons emitted per second.
(b) The intensity will have no effect on the maximum KE of emitted photoelectrons as each photon has the same energy and the 'surplus' energy (what remains of the photon energy having 'paid the energy price' to liberate an electron – the work function) will cause all electrons freed from the metal surface to have the same maximum KE. (Electrons further from the surface require more energy to free them from the metal so will not be emitted at this threshold frequency.)

56. Line spectra and the eV

1 The smallest energy transition to the ground state is from energy level 2, $E_2 - E_1$. The transition from -3.40 eV to -13.6 eV results in the emission of a photon with associated wavelength of 122 nm. This was shown in the worked example. This is shorter than the shortest visible spectrum wavelength of 400 nm. All other possible transitions from excited states to the ground state are more energetic and therefore produce photons of even shorter wavelengths.

2 The energy of a photon with wavelength 434 nm is given by
$E = hf$ where $f = \frac{c}{\lambda}$
so $f = \frac{3.00 \times 10^8}{(434 \times 10^{-9})} = 6.91 \times 10^{14}$ Hz
$E = 6.63 \times 10^{-34} \times 6.91 \times 10^{14} = 4.58 \times 10^{-19}$ J.
To convert this to eV, divide by 1.60×10^{-19} J eV^{-1} $\rightarrow 2.86$ eV
$E_5 =$ energy of emitted photon $+ E_2$
so $E_5 = 2.86$ eV $+ (-3.40$ eV$) \rightarrow E_5 = -0.54$ eV

58. Impulse and change of momentum

(a) 5.0 kg m s^{-1}
(b) $F\Delta t = 5.0$ N s so $F = \frac{5.0}{2.0} = 2.5$ N
(c) The rocket starts from rest, so its theoretical maximum velocity upwards depends on the impulse.
$p = mv = 5.0$, so $v = \frac{5.0}{0.100} = 50$ m s^{-1}
(d) Frictional forces would tend to reduce the final velocity. In addition, the mass of the rocket would decrease as fuel is burnt. This would tend to increase the final velocity.

59. Conservation of momentum in two dimentions

1 (a) $1400 \times 25 \cos(20°) = 33\,000 \text{ kg m s}^{-1}$
   (b) $1400 \times 25 \sin(20°) = 12\,000 \text{ kg m s}^{-1}$ (2 s.f.)

2 kinetic energy E_K before $= E_K$ after
   $\frac{1}{2}m_A u^2 = \frac{1}{2}m_A v_A{}^2 + \frac{1}{2}m_B v_B{}^2$

60. Elastic and inelastic collisions

1 (a) After the collision, the total linear momentum of the system is zero. Momentum must be conserved in a collision, if there are no external resultant forces, so the momentum of each rugby player before the collision must have the same magnitude but in opposite directions. Momentum is a vector.
   (b) Both players have kinetic energy before the collision. Kinetic energy is a scalar quantity, so there is a lot of kinetic energy in the system before they collide. After the collision there is none, so kinetic energy has not been conserved. This is an inelastic collision.

2 (a) Momentum is conserved: $1200 \times 10 = 4800v$
   so $v = \frac{12\,000}{4800} = 2.5 \text{ m s}^{-1}$
   (b) E_K before $= \frac{1}{2} \times 1200 \times 10^2 = 60\,000 \text{ J}$
   E_K after $= \frac{1}{2} \times 4800 \times 2.5^2 = 15\,000 \text{ J}$
   $60\,000 - 15\,000 = 45\,000 \text{ J}$ of energy was transferred away in the collision.

61. Investigating momentum change

(a) Images are equally spaced so it covers an equal distance in each equal time interval and since it is also moving in a straight line it has constant velocity.
(b) Spacing of images is greater for the red ball then the blue ball so it covers more distance in each interval of time and so is moving faster.
(c) By using the conservation of momentum along the x and y axes: speed of red ball = 1.30 ms^1 and speed of blue ball = 0.75 ms^1

63. Describing rotational motion

1 (a) $T = \frac{2\pi}{\omega}$, so $\omega = \frac{2\pi}{T}$
   $T = \frac{1}{3.0}$
   $\omega = 2\pi \times 3.0 = 19 \text{ rad s}^{-1}$
   (b) $v = \omega r = 19 \times 0.90 = 17 \text{ m s}^{-1}$

2 $\omega = \frac{\Delta\theta}{\Delta t} = \frac{45 \times 2\pi}{60} = 4.7 \text{ rad s}^{-1}$

3 (a) $\omega = \frac{2\pi}{(24 \times 3600)} = 7.3 \times 10^{-5} \text{ rad s}^{-1}$. This is exactly the same as in the worked example because all points on the Earth rotate with the same angular velocity.
   (b) The radius of the circle at latitude 51° N is $6400 \cos 51°$
   = 4027 km
   $v = \omega r = 7.3 \times 10^{-5} \times 4027 \times 10^3 = 294 \text{ m}^{-1}$

64. Uniform circular motion

1 $a = r\omega^2$
   $\omega = \frac{\theta}{t} = \frac{2\pi}{T}$
   $a = r\left(\frac{2\pi}{T}\right)^2$
   $a = 3.8 \times 10^8 \times \left(\frac{2\pi}{27.3 \times 24 \times 3600}\right)^2$
   $a = 2.7 \times 10^{-3} \text{ m s}^{-2}$ toward the centre of the Earth

2 At A: velocity = 0.
   At B: velocity = $2v$ in the direction of travel.

65. Centripetal force and acceleration

1 (a) At the minimum speed $N = 0$. Below this speed there is no contact force and the weight is now greater than the required centripetal force. This causes the car to move in a circle of radius smaller than r, so it falls away from the track.
   (b) This will happen at the speed at which $N = 0$. At this speed $\frac{mv^2}{r} = mg$, so $v = \sqrt{rg}$
   (c) At B the car is a distance $h - 2r$ below its release height and it has a speed $v = \sqrt{rg}$
   From conservation of energy, the car's kinetic energy must be equal to its loss in gravitational potential energy in descending from height h to height $2r$:
   $\frac{1}{2}mv^2 = \frac{1}{2}mrg = mg(h - 2r)$, so $h = \frac{5}{2}r$

2 The maximum value for ω is 160 rad s^{-1}
   $F = mr\omega^2 = 0.0050 \times 10^{-6} \times 0.02 \times 160^2 = 2.6 \times 10^{-6} \text{ N}$

66. Electric field strength

1 The gold nucleus creates an electric field in the space around it. The alpha particle experiences a force from this field ($F = EQ$). The alpha particle and the nucleus are both positively charged, so the force is a repulsion.

2 (a) $E = \frac{V}{d} = \frac{100}{0.05} = 2000 \text{ V m}^{-1}$ (or N C^{-1})
   (b) $F = EQ = 2000 \times 3.2 \times 10^{-19} = 6.4 \times 10^{-16} \text{ N}$

67. Electric field and electric potential

(a) $E = \frac{V}{d} = 4375 \text{ V m}^{-1}$

(b) $a = \frac{F}{m} = \frac{EQ}{m} = 1.8 \times 10^{10} \text{ m s}^{-2}$

(c) $v = \sqrt{\left(\frac{2QV}{m}\right)} = 54\,300 \text{ m s}^{-1}$

68. Forces between charges

1 (a) $F = \frac{Q_1 Q_2}{4\pi\varepsilon_0 r^2} = 0.0225 \text{ N}$ toward the 50 nC charge along the line joining their centres.
   (b) 0.0225 N toward the 20 nC charge along the line joining their centres. (This is an example of Newton's third law.)

2 $F = \frac{Q_1 Q_2}{4\pi\varepsilon_0 r^2}$
   $= \frac{(2 \times 1.60 \times 10^{-19}) \times (79 \times 1.60 \times 10^{-19})}{(4 \times \pi \times 8.85 \times 10^{-12} \times 10^{-12} \times 10^{-12})}$
   $= 3.64 \times 10^{-2} \text{ N}$

69. Field and potential for a point charge

1 $V = \frac{Q}{(4\pi\varepsilon_0 r)} = 18 \text{ kV}$
   $E = \frac{Q}{(4\pi\varepsilon_0 r^2)} = 7.2 \times 10^5 \text{ V m}^{-1}$ directed radially away from the 50 nC charge.

2 Electric field strength from one charge is
   $E = \frac{Q}{(4\pi\varepsilon_0 r^2)} = 2.3 \times 10^{10} \text{ V m}^{-1}$, so resultant field strength at the mid-point $= 4.6 \times 10^{10} \text{ V m}^{-1}$ directed toward the negative charge.
   $V = 0$. Equal magnitudes but opposite signs from each charge.

70. Capacitance

(a) $C = \frac{Q}{V} = \frac{160 \times 10^{-6}}{8.0} = 20 \text{ μF}$

(b) $Q = CV = 20 \times 10^{-6} \times 12 = 240 \text{ μC}$

(c) $I = \frac{V}{R} = \frac{12}{50} = 0.24 \text{ A} = 240 \text{ mA}$

71. Energy stored by a capacitor

1 The charging efficiency is independent of the circuit resistance. The efficiency is the same whatever the circuit resistance.

2 (a) $W = \frac{1}{2}\frac{Q^2}{C} = 1.8 \times 10^{-3}\,\text{J}$

(b) $P = \frac{W}{t} = 0.18\,\text{W}$

(c) The discharge current is high to start with but decays as the capacitor discharges. The power supplied to the resistor is $P = I^2 R$; as I falls, so does P.

72. Charging and discharging capacitors

initial current $I_0 = \frac{V}{R} = \frac{6.0}{4000} = 0.0015\,\text{A}$

Current decreases to $0.0015\,\text{A} \times 0.5$ in 139 s. Therefore, it decreases to $0.0015\,\text{A} \times (0.5)^2$, one-quarter, in $139\,\text{s} \times 2 = 278\,\text{s}$ and to $0.0015\,\text{A} \times (0.5)^3$, one-eighth, in $139\,\text{s} \times 3 = 417\,\text{s}$

73. The time constant

1 (a) For complete charging, $5RC$ must be no more than 1.0 s

$5RC \leqslant 1.0$, so $R \leqslant \frac{1.0}{5C} = 910\,\Omega$

(b) If the resistance is larger, the charging current will be lower, so the time to charge will be greater than 1.0 s.

2 $R = \frac{V}{I}$ so $1\,\Omega = 1\,\text{V A}^{-1}$, and $C = \frac{Q}{V}$ so $1\,\text{F} = 1\,\text{C V}^{-1}$

Units of RC are therefore $\text{V A}^{-1} \times \text{C V}^{-1} = \text{C A}^{-1}$ and $1\,\text{A} = 1\,\text{C s}^{-1}$ so $1\,\text{C A}^{-1} = 1\,\text{s}$.
The unit is the second.

74. Exponential decay of charge

1

$Q = Q_0 e^{-t/RC}$	$Q = Q_0 e^{-1}$	$Q = Q_0 e^{-2}$	$Q = Q_0 e^{-3}$	$Q = Q_0 e^{-4}$	$Q = Q_0 e^{-5}$
for $RC = $ 0.0235 s	$Q = 0.37 Q_0$	$Q = 0.14 Q_0$	$Q = 0.05 Q_0$	$Q = 0.02 Q_0$	$Q < 0.01 Q_0$
for $RC = $ 0.0094 s	$Q = 0.37 Q_0$	$Q = 0.14 Q_0$	$Q = 0.05 Q_0$	$Q = 0.02 Q_0$	$Q < 0.01 Q_0$

The new capacitor discharge time is $\frac{200}{500}$ times of that with the larger resistance, because charge flows faster from the capacitor with lower resistance in the circuit, but the discharge time as a function of the time constant remains the same and the capacitor is fully discharged after $5RC$.

2 (a) $Q_0 = CV_0 = 1.75 \times 10^{-6}\,\text{C}$

(b) $RC = 500 \times 10^{-9} \times 220 = 1.1 \times 10^{-4}\,\text{s} = 110\,\mu\text{s}$

(c) $Q = Q_0 e^{-t/RC} = 1.1 \times 10^{-6}\,\text{C}$

3 $e^{-t/RC} = 0.5$ so $-t/RC = \ln(0.5)$ and $t = RC\ln 2$

76. Describing magnetic fields

(a) $\Phi = BA \sin\theta = 40 \times 10^{-6} \times 1.0 \times \sin(50°) = 3.1 \times 10^{-5}\,\text{Wb}$

(b) (i) Hold it with the axis of the coil parallel to the field direction.

(ii) $N\Phi = NBA = 50 \times 40 \times 10^{-6} \times (0.30)^2$
$= 1.8 \times 10^{-4}\,\text{Wb turns}$

(iii) $\frac{1.8 \times 10^{-4}}{2.0} = 9.0 \times 10^{-5}\,\text{Wb s}^{-1}$

77. Forces on moving charges in a magnetic field

1 $F = BIl = 35 \times 10^{-6} \times 5.0 \times 10\,000 \times \sin 70° = 1.6\,\text{N}$

2 Work done is force multiplied by distance moved in the direction of the force. Magnetic forces are perpendicular to velocity so the distance moved in the direction of the magnetic force is always zero.

3 Zero. The charge is moving parallel to the field lines and the magnetic force is caused by the perpendicular component of the field.

4 In all three cases the path is an arc of a circle with the particle moving down through the field. In (a) the radius is the same as for $+q$, in (b) the radius is larger and in (c) the radius is smaller.

78. Electromagnetic induction – relative motion

(a) The first peak occurs as the magnet approaches and enters the coil. The second peak occurs as the magnet leaves and moves away from the coil.

(b) The direction of flux cutting has changed. When the magnet approaches it is increasing the downward flux through the coil; as it leaves it is reducing the downward flux through the coil.

(c) It is lower because the magnet is moving more slowly as it enters the coil than when it leaves. This results in a smaller rate of change of flux linkage and so a lower induced e.m.f. Gravity causes the magnet to accelerate so it is travelling faster as it leaves and the rate of change of flux linkage is greater, resulting in a larger induced e.m.f. over a smaller time as it leaves.

79. Changing flux linkage

1 When the current is suddenly interrupted, the flux in the coil falls to zero in a very short time. This creates a large rate of change of flux linkage that induces a large e.m.f. across the coil.

2 (a) The changing flux linkage in the core will induce an e.m.f. in the core. Since iron is a conductor, this will make currents flow in the core, or eddy currents. The core also has resistance so the currents will have a heating effect and will dissipate energy.

(b) By constructing the core in this way, the current paths are interrupted and the effective resistance of the core is increased hugely. This results in smaller currents and less heating.

80. Faraday's and Lenz's laws

1 (a) Initially: $NBA = 40 \times 0.075 \times 0.016 = 0.048\,\text{Wb}$; when parallel to the field, zero.

(b) $\frac{\Delta N\Phi}{\Delta t} = \frac{0.048}{5.0} = 0.0096\,\text{Wb s}^{-1}$

(c) $\varepsilon = 0.0096\,\text{V} = 9.6\,\text{mV}$

(d) $I = \frac{\varepsilon}{R} = \frac{0.0096}{8.0} = 1.2 \times 10^{-3}\,\text{A} = 1.2\,\text{mA}$

2 $\varepsilon = \frac{-\text{d}(N\Phi)}{\text{d}t} = -(50 \times 0.050 \times 300)\cos(300 \times t)$
$= -750\cos(300 \times t)$. The peak e.m.f. is therefore 750 V

81. Alternating currents

(a) $V_{\text{rms}} = \frac{16.0}{\sqrt{2}} = 11.3\,\text{V}$

(b) $f = \frac{1}{T} = \frac{1}{0.050} = 20\,\text{Hz}$

(c) $I_{\text{rms}} = \frac{V_{\text{rms}}}{R} = \frac{11.3}{50} = 0.23\,\text{A}$

(d) $P = I_{\text{rms}}V_{\text{rms}} = 2.6\,\text{W}$

83. The Rutherford scattering experiment

1 (a) Most alpha particles undergo very small deflections, so they clearly do not pass close to a nucleus even though the gold foil is hundreds of atoms thick.

(b) If the nucleus did not have much mass it would be scattered by the alpha particles and none of the alpha particles would bounce back.

(c) Alpha particles are charged and the way they scattered from the nucleus was successfully explained by assuming they were deflected by an electrostatic force obeying Coulomb's law.

2 If nuclear radii are about 10^4 times smaller than atomic radii, then their volumes are about 10^{12} times smaller. Since most of the mass of an atom is in its nucleus, the density of nuclear matter must be about $5000 \times (10^{12}) = 5 \times 10^{15}\,\text{kg m}^{-3}$

84. Nuclear notation

1 (a) 234
   (b) 90
   (c) 144
   (d) 90

2 Both isotopes are of the same element, so they have the same number of orbiting electrons in the same configuration. This means that their chemical properties will be identical.

3 $^{14}_{7}\text{N}$, $^{15}_{7}\text{N}$, $^{39}_{19}\text{K}$, $^{41}_{19}\text{K}$

85. Electron guns and linear accelerators

(a) There is an electric field between the cathode and anode. T1his exerts a force $F = Ee$ on the electrons and produces an acceleration $a = \dfrac{Ee}{m}$.

(b) $v = \sqrt{\left(\dfrac{2eV}{m}\right)} = \sqrt{\left(\dfrac{2 \times 1.60 \times 10^{-19} \times 400}{9.11 \times 10^{-31}}\right)} = 1.2 \times 10^7\,\text{m s}^{-1}$

(c) Kinetic energy is directly proportional to accelerating voltage; $KE = eV$ so the KE is increased by a factor of 2.

(d) v is proportional to $\sqrt{V}$ and momentum is mv so the momentum is increased by a factor of $\sqrt{2}$.

86. Cyclotrons

1 If they are not moving, they will remain at rest. Magnetic fields can deflect moving charged particles but they cannot accelerate particles at rest.

2 From the equation $r = \dfrac{p}{BQ} = \dfrac{mv}{BQ}$ we can see that if B, Q and m are fixed then the maximum value of v will be directly proportional to r.

87. Particle detectors

1 A and B must have opposite charges since they curve in opposite directions.
   A has a larger radius of curvature. This might be because of a larger mass or a smaller charge than B.

2 The radius of the curved path followed by the particle is decreasing. (1 mark for the observation) As the radius is proportional to the momentum of the particle this shows that the momentum of the particle is decreasing. (1 mark for explaining what the observation means) This is because the particle is losing energy as it creates ions along its path in the bubble chamber. (1 mark for the reason)

88. Matter and antimatter

1 (a) $\Delta E = \Delta mc^2 = 2.0 \times (3.00 \times 10^8)^2 = 1.8 \times 10^{17}\,\text{J}$
   (b) Energy conversion is 33% efficient, so electrical energy transferred = $1.8 \times 10^{17} \times 0.33\,\text{J}$.
   $t = \dfrac{E}{P} = \dfrac{1.8 \times 10^{17} \times 0.33}{1 \times 10^9} = 6 \times 10^7\,\text{s} = 1.9\,\text{years}$

2 (a) If only one photon were emitted it would have linear momentum in the direction of its emission. If a pair of photons is created moving in opposite directions, momentum is conserved.
   (b) The electron and positron would carry away the excess energy as kinetic energy.

89. The structure of nucleons

Proton structure is uud: $\frac{2}{3}e + \frac{2}{3}e - \frac{1}{3}e = +e$

Neutron structure is udd: $\frac{2}{3}e - \frac{1}{3}e - \frac{1}{3}e = 0$

pi$^+$ meson structure is u$\bar{\text{d}}$: $\frac{2}{3}e + \frac{1}{3}e = +e$

pi$^-$ meson structure is d$\bar{\text{u}}$: $-\frac{1}{3}e - \frac{2}{3}e = -e$

90. Nuclear energy units

1 Rest energy of proton = $mc^2 = 1.67 \times 10^{-27} \times 9.00 \times 10^{16}$
   $= 1.50 \times 10^{-10}\,\text{J} = 9.39 \times 10^8\,\text{eV} = 939\,\text{MeV}$. So the mass in
   $\dfrac{\text{MeV}}{c^2} = 939\,\dfrac{\text{MeV}}{c^2} \approx 1\,\dfrac{\text{GeV}}{c\text{M}^2}$

2 The rest mass of an electron is $\dfrac{0.511\,\text{MeV}}{c^2}$, so its rest energy is 0.511 MeV. To give it this much kinetic energy, it must be accelerated through a p.d. of 0.511 MV.

3 $v = \sqrt{\left(\dfrac{2(E_k)}{m}\right)} = \sqrt{\left(\dfrac{2 \times 50 \times 1.6 \times 10^{-19}}{1.67 \times 10^{-27}}\right)} = 9.8 \times 10^4\,\text{m s}^{-1}$

91. The Standard Model

1 (a) $+e$
   (b) $+e$

2 Leptons are not composed of other simpler particles and have no internal structure. Baryons have an internal structure. They are composed of quarks, so cannot be said to be fundamental.

92. Particle interactions

1 (a) left-hand side charge = 1 + 1 = 2
       right-hand side charge = 1 + 1 + 1 − 1 = 2
       Therefore charge is conserved.
   (b) left-hand side baryon number = 1 + 1 = 2
       right-hand side baryon number = 1 + 1 + 0 − 1 = 1.
       Pions are mesons and so have a baryon number of zero.
       Therefore baryon number is not conserved.
   (c) Baryon number is not conserved and therefore the reaction is forbidden.

2 Electrons have lepton number +1 and antielectrons have lepton number −1. Before the annihilation, the lepton number is zero (1 − 1). After the annihilation, there are just two gamma rays. These have lepton number zero, so the lepton number is conserved in this reaction.

94. Specific heat capacity

1 $\Delta E = mc\Delta\theta = 0.800 \times 4200 \times 60 = 201\,600\,\text{J}$
   $t = \dfrac{E}{P} = \dfrac{201\,600}{50} = 4032\,\text{s} = 67.2\,\text{minutes.}$

2 Energy supplied in 1 minute = $10\,000 \times 60 = 600\,000\,\text{J}$
   $\Delta\theta = \dfrac{\Delta E}{mc} = \dfrac{600\,000}{(7.0 \times 4200)} = 20.4°\text{C}$. Therefore the water comes out at 40.4°C

95. Latent heats

(a) $L = \dfrac{IVt}{\Delta m} = \dfrac{2.5 \times 12.0 \times 8.0 \times 60}{0.0060} = 2\,400\,000\,\text{J kg}^{-1}$

(b) This result is larger than the expected value, probably because some energy was transferred to the surroundings during the experiment so not all the calculated energy was transferred to the water.

96. Pressure and volume of an ideal gas

1 (a) $3.6 \times 10^5\,\text{Pa}$
   (b) It would be greater than $3.6 \times 10^5\,\text{Pa}$ because the molecules would be moving faster and this would increase the pressure in addition to the effect of reducing the volume.

(c) It would be lower than 3.6×10^5 Pa because there would be fewer molecules to create the pressure.

2   If the volume is reduced, each molecule will collide with the walls of the container more frequently, so the force on the walls will increase and so will the pressure.

97. Absolute zero

1   $T = \theta + 273 = 310$ K

2   (a) 323 K and 393 K

(b) 70°C or 70 K

(c) Pressure is directly proportional to temperature on the Kelvin scale, so
$$p_{final} = p_{initial} \times \frac{T_{final}}{T_{initial}} = 1.01 \times 10^5 \times \left(\frac{393}{323}\right) = 1.23 \times 10^5 \text{ Pa}$$

98. Kinetic theory

(a) $pV = NkT$ so $N = \frac{pV}{kT} = \frac{(2.0 \times 10^5 \times 0.0040)}{(1.38 \times 10^{-23} \times 323)} = 1.8 \times 10^{23}$

(b) $\frac{pV}{T}$ = constant; in this case, V is a constant, so $\frac{p_2}{T_2} = \frac{p_1}{T_1}$
so $p_2 = \frac{p_1 T_2}{T_1} = 2.2 \times 10^5$ Pa

99. Particles and energy

1   $\sqrt{<c^2>} = \sqrt{\left(\frac{3kT}{m}\right)} = \sqrt{\left(\frac{3 \times 1.38 \times 10^{-23} \times 293 \times 6.02 \times 10^{23}}{32 \times 10^{-3}}\right)}$
$= 478 \text{ m s}^{-1}$

2   More of the molecules have higher kinetic energy so the collisions are more energetic. If there is a minimum (activation) energy needed to cause the reaction then more molecules reach this energy and the reaction rate increases.

3   (a) 341 m s^{-1}
(b) $116\,694.6 \text{ m}^2 \text{ s}^{-2}$
(c) 342 m s^{-1}

100. Black body radiation

(a) $\lambda_{max} = \frac{2.9 \times 10^{-3}}{5800} = 5.0 \times 10^{-7}$ m, visible

(b) $L = \sigma A T^4 = 5.67 \times 10^{-8} \times 4 \times \pi \times (7.0 \times 10^8)^2 \times 5800^4$
$= 3.95 \times 10^{26}$ W

101. Standard candles

$I = \frac{L}{4\pi d^2} = \frac{(5.6 \times 10^{31})}{(4 \times \pi \times (643 \times 3600 \times 24 \times 365 \times 3.00 \times 10^8)^2)}$
$= 1.2 \times 10^{-7}$ W

102. Trigonometric parallax

1   $d = \frac{r}{\theta} = \frac{1.5 \times 10^{11}}{10^{-7}} = 1.5 \times 10^{18}$ m

2   Convert the angle to radians:
$3 \times 10^{-5\circ} = 3 \times 10^{-5} \times \frac{\pi}{180} = 5 \times 10^{-7}$ radians
$d = \frac{r}{\theta} = \frac{1.5 \times 10^{11}}{(5.2 \times 10^{-7})} = 2.9 \times 10^{17}$ m

103. The Hertzsprung–Russell diagram

(a) as in diagram on page 103. The Sun is found on the main sequence at a surface temperature of about 6000 K. A star with lower surface temperature than the Sun will be found on the main sequence to the right of the Sun's position (remember that the temperature scale on an HR diagram is reversed).

(b) (i) in the supergiant region
(ii) in the red giant region
(iii) in the white dwarf section

104. Stellar life cycles and the Hertzsprung–Russell diagram

1   Larger stars have more mass so their cores are denser, meaning that nuclear fusion reactions proceed at a much greater rate, so they use up their nuclear fuel more rapidly.

2

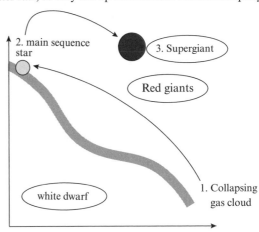

3   Its mass is too low. The pressure and temperature at its core did not reach the levels needed for nuclear fusion reactions to begin.

105. The Doppler effect

1   $\Delta\lambda = \lambda' - \lambda = 595 - 589 = 6$ nm
$z = \frac{\Delta\lambda}{\lambda} = \frac{6}{589} = 0.010 = \frac{v}{c}$
Therefore $v = 0.010 \times c = 3.0 \times 10^6 \text{ m s}^{-1}$ (moving away).

2   $d = \frac{v}{H_0} = \frac{zc}{H_0} = \frac{0.050 \times 3.00 \times 10^8}{(2 \times 10^{-18})} = 7.5 \times 10^{24}$ m away.

106. Cosmology

The more mass and energy that is present in the Universe, the greater the gravitational forces between galaxies and the more likely they are to slow in their expansion.

108. Mass and energy

1   $\Delta m = 2 \times m_e = 2 \times 9.11 \times 10^{-31}$ kg
$\Delta E = c^2 \Delta m = 9.00 \times 10^{16} \times 2 \times 9.11 \times 10^{-31} = 1.64 \times 10^{-13}$ J
For one of the photon pair, $E = hf = \frac{hc}{\lambda}$
$\lambda = \frac{hc}{E} = \frac{6.63 \times 10^{-34} \times 3.00 \times 10^8}{(0.5 \times 1.64 \times 10^{-13})} = 2.4 \times 10^{-12}$ m

2   $\Delta E = c^2 \Delta m = 9.00 \times 10^{16} \times 55.935 \times 1.66 \times 10^{-27}$
$= 8.36 \times 10^{-9}$ J
This is $\frac{8.36 \times 10^{-9}}{(1.60 \times 10^{-19})} = 5.22 \times 10^{10}$ eV = 52.2 GeV.

109. Nuclear binding energy

1   $\frac{493}{56} = 8.8$ MeV/nucleon per nucleon

2   (a) $\Delta m = 1.9561$ u $= 3.2470928 \times 10^{-27}$ kg.
Total B.E. $= c^2 \Delta m = 2.92 \times 10^{-10}$ J = 1830 MeV
(b) $\frac{\text{B.E.}}{A} = 7.7$ MeV

110. Nuclear fission

1   Energy input $= \frac{1}{0.4} = 2.5$ GW
$t = \frac{E}{P} = \frac{7.4 \times 10^{13}}{(2.5 \times 10^9)} = 29\,600$ s, or about 8.2 hours

2   Unlike ^{235}U, ^{238}U is not fissile – it does not split spontaneously when hit by a neutron. So although ^{235}U atoms in the sample may split and release neutrons, these neutrons are far more likely to hit one of the much more common ^{238}U atoms, where they will not cause immediate fission, than to hit another fissile ^{235}U atom.

111. Nuclear fusion

1. Total energy released is: $1.2 \times 10^{26} \times 2.82 \times 10^{-12}$ J $= 3.4 \times 10^{14}$ J
2. High temperatures are needed to create plasma: that is, to strip the atoms of electrons and to give nuclei sufficient kinetic energy to overcome their mutual electrostatic repulsion (since they are all positively charged). Under these conditions they can come close enough for the strong nuclear force to bind them together.

> The electron masses cancel in this reaction so it is fine to use atomic masses.

112. Background radiation

1. (a) Radioactivity is a random process so the rate of decay varies randomly.
   (b) 29 cpm (2 s.f.)
   (c) 84 cpm (2 s.f.)
   (d) 55 cpm (2 s.f.)
2. (a) Because a basement or a mine is surrounded by rock and poorly ventilated, it will gather more radon than a house or the open air. In addition, radon atoms are heavy, so radon is a dense gas and tends to sink in air.
   (b) Because radon is a gas, it can be inhaled and may become trapped in the lungs, where it delivers energy to lung cells when it decays. Other parts of the body receive lower doses from radon in the air.

113. Alpha, beta and gamma radiation

1. Most alpha particles will be stopped by the dead outer layers of skin so they do not affect living and dividing cells when the source is outside the body. However, if the source is taken inside the body then the alpha particles can cause damage inside living and dividing cells.
2. One way to do this would be to shield the source with a few millimetres of aluminium. This would absorb the beta particles but allow most of the gamma radiation to pass through.
3. Alpha particles are more strongly ionising, so they transfer energy rapidly to other particles in the air as they pass through it, ionising them and soon stopping. Beta and gamma radiation transfer energy more slowly as they are less ionising, so they travel farther.

114. Investigating the absorption of gamma radiation by lead

1. Lead is denser than steel or concrete, so there is more matter in the way of the gamma rays. This increases the probability that they will be absorbed.
2. If a source with a very short half-life were used the count rate would fall noticeably over the course of the experiment, so it would not be a fair test.

115. Nuclear transformation equations

1. $^{222}_{86}\text{Rn}$ and $^{4}_{2}\alpha$
2. (a) The electron is created in the decay. This would increase the lepton number in the Universe by 1. To balance this it follows that a particle with lepton number -1 must also be created. Baryon number and charge are already conserved without this additional particle, so it must be an uncharged antilepton, an antineutrino.
   (b) $^{1}_{0}n \rightarrow\, ^{1}_{1}p +\, ^{0}_{-1}e +\, ^{0}_{0}\bar{v}$
   Baryon number: $1 = 1 + 0 + 0$
   Charge: $0 = 1 - 1 + 0$
   Lepton number: $0 = 0 + 1 - 1$

116. Radioactive decay and half-life

(a) $\frac{1}{4} = \frac{1}{2^2}$ so two half-lives have passed. Age $\approx 2 \times 5700$ = 11 400 years.
(b) $\frac{1}{16} = \frac{1}{2^4}$ so four half-lives have passed. Age $\approx 4 \times 5700$ = 22 800 years.
(c) Rocks were never alive.

117. Exponential decay

$\frac{t_1}{2} = 185$ s.

118. Radioactive decay calculations

1. $\lambda = 3.74 \times 10^{-3}$ s^{-1} and $t_{\frac{1}{2}} = 185$ s
2. $t_{\frac{1}{2}} = \dfrac{\ln 2}{\lambda} = \dfrac{\ln 2}{(5.1 \times 10^{-11})} = 1.4 \times 10^{10}$ s $= 430$ years
   For $A_0 = 1$, $A = 0.05$
   $A = A_0 e^{-\lambda t}$
   $0.05 = e^{-\lambda t}$
   $\ln 0.05 = -5.1 \times 10^{-11} \times t$
   $t = 5.9 \times 10^{10}$ s $= 1900$ years

119. Gravitational fields

1. 590 N, 98 N
2. Although the field lines at the Earth's surface do diverge with distance from the surface, on the scale at which we are measuring them they are approximately parallel and equally spaced. The gravitational field strength g is constant at the level we use, of three significant figures.

120. Gravitational potential and gravitational potential energy

1. (a) 15000 J (ii) 2400 J
   (b) (i) 250 J kg^{-1} (ii) 41 J kg^{-1}
2. The force on a mass in a uniform gravitational field is constant, so the work done to lift it through any constant distance h is always the same. This means that equal increases in potential occur in equal distances, so the equipotentials are equally spaced.

121. Newton's law of gravitation

(a) $F = \dfrac{Gm_1m_2}{r_E^2} = \dfrac{6.67 \times 10^{-11} \times 60 \times 6.0 \times 10^{24}}{(6400 \times 10^3)^2} = 586$ N
(b) g = gravitational force per unit mass exerted on a mass at the surface of the Earth.
   $g = \dfrac{F}{m}$
(c) $g = \dfrac{586}{60} = 9.8$ N kg^{-1}

122. Gravitational field of a point mass

1. $g = \dfrac{Gm}{r^2} = \dfrac{6.67 \times 10^{-11} \times 6.0 \times 10^{24}}{(380\,000\,000)^2} = 2.8 \times 10^{-3}$ N kg^{-1}
2. At the surface, distance R, $g_R = \dfrac{GM}{R^2} = 1.6$
   at R from the surface, $g_{2R} = \dfrac{GM}{(2R)^2} = \dfrac{1.6}{(2)^2} = 0.40$ Nkg^{-1}
   and at 4R from the surface, $g_{5R} = \dfrac{GM}{(5R)^2} = \dfrac{1.5}{(5)^2} = 0.064$ Nkg^{-1}
3. A small mass at the centre of the Earth will be attracted to all of the matter inside the Earth. This mass is distributed uniformly around it so the nett force will be zero, and so will the field strength.

123. Gravitational potential in a radial field

(a) $V_{grav} = \dfrac{-GM}{r} = \dfrac{-6.67 \times 10^{-11} \times 6.0 \times 10^{24}}{(6.4 \times 10^6 + 0.59 \times 10^6)} = -5.7 \times 10^7 \, \text{J kg}^{-1}$

(b) $W = m\Delta V_{grav} = 10\,000 \, (6.3 \times 10^7 - 5.7 \times 10^7) = 6.0 \times 10^{10} \, \text{J}$

(c) This does not take into account the mass of the launch vehicle, the mass of the fuel that must be lifted, losses to air resistance etc., and the fact that the telescope also needs kinetic energy to orbit.

124. Energy changes in a gravitational field

(a) $+57 \, \text{MJ}$

(b) It must be given $KE = 57 \, \text{MJ}$, therefore, since $\frac{1}{2}mv^2 = 57 \, \text{MJ}$,

$v = \sqrt{\left(\dfrac{2 \times 57 \times 10^6}{1.0}\right)} = 10\,700 \, \text{m s}^{-1}$

(c) For a mass m, both the kinetic energy supplied and the change in GPE required are directly proportional to m, so it cancels.

125. Comparing electric and gravitational fields

1   It would remain the same because both forces obey an inverse square law. If the distance between the particles is increased by a factor of 10 then both forces are reduced by a factor of 10^2 (100) so their ratio is unaffected.

2   In a gravitational field $a = \dfrac{F_{grav}}{m} = \dfrac{mg}{m} = g$ (m cancels).

In an electric field $a = \dfrac{F_{electric}}{m} = \dfrac{QE}{m}$. This depends on the

ratio of $\dfrac{Q}{m}$ so varies for different charged particles.

126. Orbits

1   $\dfrac{Gm_1m_2}{r^2} = \dfrac{m_2v^2}{r}$

$\dfrac{Gm_1}{r^2} = \dfrac{4\pi^2 r^2}{rT^2}$

$r^3 = \dfrac{Gm_1T^2}{4\pi^2} = 7.57 \times 10^{22} \, \text{m}^3$

r = 42 000 km (this is the radius of the orbit, not the height above Earth's surface)

2   The axis of rotation of the satellite is different from the axis of rotation of the Earth so whilst it will be over the same point once every 24 hours, the Earth will rotate beneath it as it turns.

128. Simple harmonic motion

(a) The fourth measurement, 10.06 s

(b) The other four results are closely grouped. This result is about 1.1 s different from those results, and 1.1 s is approximately equal to the time period of the oscillator. The student probably miscounted the number of oscillations (9 instead of 10).

(c) The anomalous result must be ignored. The average time for one oscillation is $\dfrac{(11.12 + 11.22 + 11.18 + 11.15)}{40} = 1.12 \, \text{s}$

129. Analysing simple harmonic motion

(a) $a = \dfrac{F}{m} = \dfrac{0.80}{0.60} = 1.3 \, \text{m s}^{-2}$

(b) $a = -\omega^2 x$ so $\omega = \sqrt{\left(\dfrac{-a}{x}\right)} = \sqrt{\left(\dfrac{1.3}{0.025}\right)} = 7.2 \, \text{s}^{-1}$

(The minus sign disappears because acceleration and displacement are in opposite directions.)

(c) $f = \dfrac{\omega}{2\pi} = \dfrac{7.211}{2\pi} = 1.15 \, \text{Hz}$

(d) $v_{max} = \omega A = 0.18 \, \text{m s}^{-1}$ at the maximum amplitude.

130. Graphs of simple harmonic motion

The graph is an inversion of the displacement–time graph because $a = -\omega^2 x$.

Magnitude of maximum acceleration $a_{max} = \omega^2 A$

$= \dfrac{4\pi^2 A}{T^2} = 1110 \, \text{m s}^{-2}$

131. The mass–spring oscillator and the simple pendulum

1   $T = 2\pi\sqrt{\left(\dfrac{l}{g}\right)}$ so $l = \dfrac{gT^2}{4\pi^2} = 0.16 \, \text{m}$

2   The period of the mass–spring pendulum does not depend on g, so it is unaffected (still 1.0 s).
The period of the simple pendulum is proportional to $\dfrac{1}{\sqrt{g}}$, so it will increase as g gets smaller (>1.0 s).

132. Energy and damping in simple harmonic oscillators

for a mass–spring oscillator, $\omega^2 = \dfrac{k}{m}$

$$\frac{1}{2}m\omega^2 A^2 = \frac{1}{2}m\left(\frac{k}{m}\right)A^2 = \frac{1}{2}kA^2$$

$$= \frac{1}{2} \times 25 \times (0.05)^2$$

total energy $= 0.031 \, \text{J}$

133. Forced oscillations and resonance

If the parts in the speaker are not properly isolated from each other, then when the music played makes the speaker diaphragm itself vibrate at the resonant frequency of the case or some other component with which it is in contact, that component will move too, generating the buzz by friction.

134. Driven oscillators

(a) The bump causes the wheel to move, compressing the spring, and the wheel will then oscillate up and down. (The effect causes the contact force on the road to vary and, therefore, the friction force also – this will cause loss of steering control.)

(b) The shock absorbers 'damp' the oscillation by dissipating the energy of the oscillation.

Periodic table

Key

Atomic (proton number)
Atomic symbol
Name
Relative atomic mass

Example: 1 **H** Hydrogen 1.0

Group → Period ↓	1 (1)	2 (2)	(3)	(4)	(5)	(6)	(7)	(8)	(9)	(10)	(11)	(12)	3 (13)	4 (14)	5 (15)	6 (16)	7 (17)	8 (18)
1	1 **H** Hydrogen 1.0																	2 **He** Helium 4.0
2	3 **Li** Lithium 6.9	4 **Be** Beryllium 9.0											5 **B** Boron 10.8	6 **C** Carbon 12.0	7 **N** Nitrogen 14.0	8 **O** Oxygen 16.0	9 **F** Fluorine 19.0	10 **Ne** Neon 20.2
3	11 **Na** Sodium 23.0	12 **Mg** Magnesium 24.3											13 **Al** Aluminium 27.0	14 **Si** Silicon 28.1	15 **P** Phosphorus 31.0	16 **S** Sulfur 32.1	17 **Cl** Chlorine 35.5	18 **Ar** Argon 39.9
4	19 **K** Potassium 39.1	20 **Ca** Calcium 40.1	21 **Sc** Scandium 45.0	22 **Ti** Titanium 47.9	23 **V** Vanadium 50.9	24 **Cr** Chromium 52.0	25 **Mn** Manganese 54.9	26 **Fe** Iron 55.8	26 **Co** Cobalt 58.9	28 **Ni** Nickel 58.7	29 **Cu** Copper 63.5	30 **Zn** Zinc 65.4	31 **Ga** Gallium 69.7	32 **Ge** Germanium 72.6	33 **As** Arsenic 74.9	34 **Se** Selenium 79.0	35 **Br** Bromine 79.9	36 **Kr** Krypton 83.8
5	37 **Rb** Rubidium 85.5	38 **Sr** Strontium 87.6	39 **Y** Yttrium 88.9	40 **Zr** Zirconium 91.2	41 **Nb** Niobium 92.9	42 **Mo** Molybdenum 95.9	43 **Tc** Technetium (98)	44 **Ru** Ruthenium 101.1	45 **Rh** Rhodium 102.9	46 **Pd** Palladium 106.4	47 **Ag** Silver 107.9	48 **Cd** Cadmium 112.4	49 **In** Indium 114.8	50 **Sn** Tin 118.7	51 **Sb** Antimony 121.8	52 **Te** Tellurium 127.6	53 **I** Iodine 126.9	54 **Xe** Xenon 131.3
6	55 **Cs** Caesium 132.9	56 **Ba** Barium 137.3	57 **La*** Lanthanum 138.9	72 **Hf** Hafnium 178.5	73 **Ta** Tantalum 180.9	74 **W** Tungsten 183.8	75 **Re** Rhenium 186.2	76 **Os** Osmium 190.2	77 **Ir** Iridium 192.2	78 **Pt** Platinum 195.1	79 **Au** Gold 197.0	80 **Hg** Mercury 200.6	81 **Tl** Thallium 204.4	82 **Pb** Lead 207.2	83 **Bi** Bismuth 209.0	84 **Po** Polonium (209)	85 **At** Astatine (210)	86 **Rn** Radon (222)
7	87 **Fr** Francium (223)	88 **Ra** Radium (226)	89 **Ac*** Actinium (227)	104 **Rf** Rutherfordium (261)	105 **Db** Dubnium (262)	106 **Sg** Seaborgium (266)	107 **Bh** Bohrium (264)	108 **Hs** Hassium (277)	109 **Mt** Meitnerium (268)	110 **Ds** Darmstadtium (271)	111 **Rg** Roentgenium (272)	112 **Cn** Copernicium 112		114 **Fl** Flerovium 114		116 **Lv** livermorium 116		

58 **Ce** Cerium 140.1	59 **Pr** Praseodymium 140.9	60 **Nd** Neodymium 144.2	61 **Pm** Promethium 144.9	62 **Sm** Samarium 150.4	63 **Eu** Europium 152.0	64 **Gd** Gadolinium 157.2	65 **Tb** Terbium 158.9	66 **Dy** Dysprosium 162.5	67 **Ho** Holmium 164.9	68 **Er** Erbium 167.3	69 **Tm** Thulium 168.9	70 **Yb** Ytterbium 173.0	71 **Lu** Lutetium 175.0
90 **Th** Thorium 232.0	91 **Pa** Protactinium (231)	92 **U** Uranium 238.1	93 **Np** Neptunium (237)	94 **Pu** Plutonium (242)	95 **Am** Americium (243)	96 **Cm** Curium (247)	97 **Bk** Berkelium (245)	98 **Cf** Californium (251)	99 **Es** Einsteinium (254)	100 **Fm** Fermium (253)	101 **Md** Mendelevium (256)	102 **No** Nobelium (254)	103 **Lr** Lawrencium (257)

Notes

Published by Pearson Education Limited, 80 Strand, London, WC2R 0RL.

www.pearsonschoolsandfecolleges.co.uk

Copies of official specifications for all Edexcel qualifications may be found on the website: www.edexcel.com

Text © Pearson 2016
Typeset and illustrated by Tech-Set Ltd, Gateshead
Produced by Out of House Publishing
Cover illustration by Miriam Sturdee
Original illustrations © Pearson Education 2016

The rights of Steve Woolley and Steve Adams to be identified as authors of this work have been asserted by them in accordance with the Copyright, Designs and Patents Act 1988.

First published 2016

19 18 17 16
10 9 8 7 6 5 4 3 2 1

British Library Cataloguing in Publication Data
A catalogue record for this book is available from the British Library

ISBN 978 1 447 98998 1

Printed in Slovakia by Neografia

Acknowledgements
The publisher would like to thank the following for their kind permission to reproduce their photographs:

(Key: b-bottom; c-centre; l-left; r-right; t-top)

Shutterstock.com: Ti Santi 19, **Alamy Images:** David Burton 60, PCN Photography 58; **CERN Geneva:** 90; **Fotolia.com:** Budimir Jevtic 30; **Pearson Education Ltd:** Trevor Clifford 19, Tsz-shan Kwok 29, 93; **Science Photo Library Ltd:** Carl Anderson 88, David Parker 89, European Space Agency 102, Library of Congress 54cr; **Shutterstock.com:** Esteban De Armas 108, Georgios Kollidas 54tr, yurazaga 70; **Wellcome Library, London:** 54cl

All other images © Pearson Education